Notes on
DIFFERENTIAL GEOMETRY

NOEL J. HICKS

Department of Mathematics
The University of Michigan

VAN NOSTRAND REINHOLD COMPANY
LONDON

NEW YORK CINCINNATI TORONTO MELBOURNE

VAN NOSTRAND REINHOLD COMPANY
25-28 Buckingham Gate, London SW1E 6LQ

INTERNATIONAL OFFICES
New York Cincinnati Toronto Melbourne

Library of Congress Catalog Card No. 77-11338

ISBN 0 442 034059

First published 1971
Reprinted 1974

Printed and bound in Great Britain by
REDWOOD BURN LIMITED
Trowbridge & Esher

PREFACE

The following paragraph presents a very brief history of differential geometry and the notation used in these notes.

Differential geometry is probably as old as any mathematical discipline and certainly was well launched after Newton and Leibnitz had laid the foundations of calculus. Many results concerning surfaces in 3-space were obtained by Gauss in the first half of the nineteenth centruy, and in 1854 Riemann laid the foundations for a more abstract approach. At the end of that century, Levi-Civitae and Ricci developed the concept of parallel translation in the classical language of tensors. This approach received a tremendous impetus from Einstein's work on relativity. During the early years of this century, E. Cartan initiated research and methods that were independent of a particular coordinate system (invariant methods). Chevalley's book "The Theory of Lie Groups" (1946) continued the clarification of concepts and notation, and it has had a remarkable affect on the current situation. The complete global synthesis of Cartan's approach was achieved when Ehresmann formulated a connexion in terms of a fiber bundle. These notes utilize an invariant local method formulated by Koszul.

The first three chapters of this book provide a short course on classical differential geometry and could be used at the junior level with a little outside reading in linear algebra and advanced calculus. The first six chapters can be used for a one-semester course in differential geometry at the senior-graduate level. Such a course would cover the main topics of classical differential geometry (except for the material in chapter 8) using modern language and techniques, and it would prepare a student for further study in the books of Helgason, Lang, Sternberg, etc. (see list in following paragraph). The entire book can be covered in a full year course. A selection of chapters could make up a "topics" course or a course on Riemannian geometry.

For example, a course on manifolds and connexions could consist of chapters 1, 4, 5, 7, and sections 9.1, 9.3, and 9.4. The reader with a little experience should move through the first three chapters fairly quickly.

The problems are of several types: (a) those that provide explicit computations to test the understanding of the theory, (b) those that require the student to prove theorems similar to those in the text, (c) those that lead the student through supplementary material, some of which may be an integral part of the exposition, and (d) those that lead the student to books or papers in the literature. An introduction to bundle theory and the theory of Lie groups is covered via problem material. Our hope is to give the reader a solid understanding of the basic concepts and to stimulate him to further reading and thinking in differential geometry.

Besides the specific references found in the notes, we would like to mention the following general references: Point set topology: Kelley; Hocking and Young; Pervin. Linear algebra: Halmos; Jacobson. Advanced calculus: Buck; Kaplan; Nickerson, Steenrod, and Spencer. Classical differential geometry: Eisenhart; Hilbert and Cohn-Vossen; Struik. Contemporary differential geometry: Auslander and MacKenzie; Crittenden and Bishop; Guggenheimer; Helgason; Kobayashi and Nomizu; Lang; Normizu; Sternberg. History of differential geometry: Struik; Veblen and Whitehead.

We will use the following conventions: "iff" for "if and only if"; "//" for "Q. E. D."; Cartan^3 will refer to the third reference in the bibliography under Cartan, and when there is only one reference for an author, we omit the superscript 1; $\Sigma_{i=1}^{n}$, Σ_i, and Σ will all be used to indicate a sum is to be made, and in the latter two cases, we hope the omitted information (range or index of summation) is clear from the context.

At this time I would like to express my gratitude to former teachers N. Schwid and V. J. Varineau for their early encouragement, to Miss Margaret M. Genova and Miss Gillian D. Hodge for their help in typing the manuscript, and to L. M. Dickens for his contribution to the understanding of the illustrations. Finally, I am indebted to W. Ambrose and H. Samelson for sharing their insights via courses, notes, and conversations.

N. J. HICKS

Ann Arbor, Michigan
May 1964

CONTENTS

1. Manifolds

In this chapter we define the fundamental concepts which we deal with throughout these notes. Specifically, the notions of manifold, function, and vector, and the concept of differentiability (smoothness), must be carefully digested for a solid foundation.

Section 1.1 *Manifolds*

First some notation. Let R be the set of real numbers. For an integer $n > 0$, let R^n be the product space of ordered n-tuples of real numbers. Thus $R^n = [(a_1,\ldots,a_n)\colon\ a_i \text{ in } R]$. For $i = 1,\ldots,n$, let u_i be the natural coordinate (slot) functions of R^n, i.e., $u_i\colon\ R^n \to R$ by $u_i(a_1,\ldots,a_n) = a_i$. An open set of R^n will be a set which is open in the standard metric topology induced by the standard metric function d on R^n; thus if $a = (a_1,\ldots,a_n)$ and $b = (b_1,\ldots,b_n)$ are points in R^n, then $d(a, b) = [\Sigma_{i=1}^n (a_i - b_i)^2]^{1/2}$.

The concept of differentiability is based ultimately on the definition of a derivative in elementary calculus. Let r be an integer, $r > 0$. Recall from advanced calculus that a map f from an open set $A \subset R^n$ into R is called C^r on A if it possesses continuous partial derivatives on A of all orders $\leq r$. If f is merely continuous from A to R, then f is C^0 on A. If f is C^r on A for all r, then f is C^∞ on A. If f is real analytic on A (expandable in a power series in the coordinate functions about each point of A), then f is C^ω on A. Henceforth, unless otherwise specified, *we let r be ∞, ω, or an integer > 0.*

A map f from an open set $A \subset R^n$ into R^k (k an integer ≥ 1) is C^r on A if each of its slot functions $f_i = u_i \circ f$ is C^r on A for $i = 1,\ldots,k$; thus for p in R^n, $f(p) = (f_1(p),\ldots,f_k(p))$ in R^k.

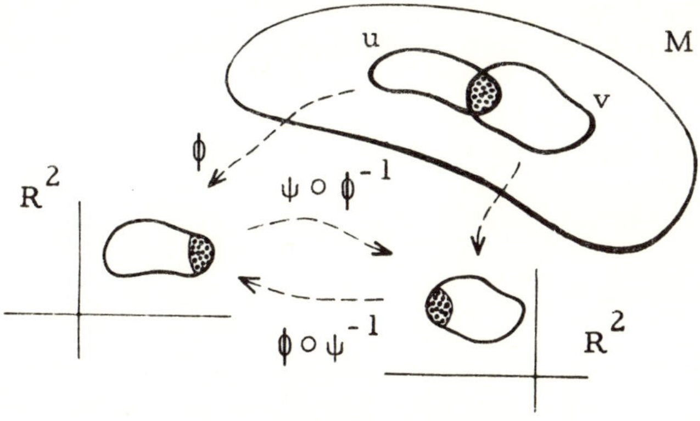

Fig. 1.1 Overlapping Coordinate Domains

We now define a manifold. Let M be a set. An *n-coordinate pair* on M is a pair (ϕ, U) consisting of a subset U of M and 1 to 1 map ϕ of U onto an open set in R^n. One n-coordinate pair (ϕ, U) is C^r *related* to another n-coordinate pair (θ, V) if the maps $\phi \circ \theta^{-1}$ and $\theta \circ \phi^{-1}$ are C^r maps wherever they are defined (thus their domains of definition must be open). A C^r *n-subatlas* on M is a collection of n-coordinate pairs (ϕ_h, U_h), each of which is C^r related to every other member of the collection, and the union of the sets U_h is M. A maximal collection of C^r related n-coordinate pairs is called a C^r *n-atlas*. If a C^r n-atlas contains a C^r n-subatlas, we say the subatl as induces or generates the atlas. Finally, an *n dimensional* C^r *manifold* or a C^r *n-manifold* is a set M together with a C^r n-atlas. When $r = 0$, M is customarily called a *locally Euclidean space* or a *topological manifold,* and only when $r \neq 0$ is M called a *differentiable* or *smooth* manifold. An atlas on a set M is often called a *differentiable structure* or a *manifold structure* on M. Notice that one set may possess more than one differentiable structure (see example 4 below), however, a definition of "equivalent" differentiable structures is necessary before the study of different atlases on a set becomes meaningful (see Munkres[1]).

Each n-coordinate pair (ϕ, U) on a set M induces a set of n real valued functions on U defined by $x_i = u_i \circ \phi$ for $i = 1,\ldots,n$. The functions $x_1,\ldots,x_n$ are called *coordinate functions* or *a coordinate system* and U is called the *domain* of the coordinate system.

We list some examples:

1. Let M be R^n with a C^r n-subatlas equal to the pair consisting of ϕ = the identity map and $U = R^n$.

2. Let M be any open set of R^n and let a $C^r n$-subatlas be (the identity map, M).

3. Let $M = GL(n, R)$, the group of non-singular R-linear transformations of R^n into itself. Then M can be mapped 1:1 onto an open set in R^{n^2} and thus a manifold structure can be defined on M via example 2. If (a_{ij}) is a matrix representation of an element of M with respect to the usual base of R^n, then map (a_{ij}) into the n^2-tuple

$$(a_{11}, a_{12},\ldots,a_{1n}, a_{21}, a_{22},\ldots,a_{2n}, a_{31},\ldots,a_{nn}).$$

The image set of this map will be open since it is the inverse image of an open set by the determinant map, which is continuous (indeed it is C^ω as a map on R^{n^2}).

4. Let M_1 be the 1-dimensional C^1 manifold of example 1, and let $M_2 = R$ with the C^1 1-subatlas (x^3, R), where x is the identity mapping on R. Then $M_1 \neq M_2$ since $x^{1/3}$ is not C^1 at the origin.

5. Let f be a C^r real valued function on R^{n+1}, with $r > 0$ and $n > 0$, and suppose the gradient of f does not vanish on an f-constant set $M = [p \text{ in } R^{n+1}\colon\ f(p) = 0]$. Then at each point in M, choose any partial derivative of f that doesn't vanish, say the ith one, apply the implicit function theorem to obtain a neighborhood of p (relative topology on M) which projects in a 1:1 way into the $u_i = 0$ hyperplane of R^{n+1}, and thus define a subatlas which makes M a C^r n-manifold.

   This example covers many classical hypersurfaces in R^{n+1}, including spheres, planes, and cylinders.

6. The process in example 5 can easily be generalized to obtain C^r $(n-k)$-manifolds from "constant sets" of a C^r map $f\colon\ R^n \to R^k$ whose Jacobian matrix is of rank k on the constant set.

7. Let F be a univalent map from an open set in R^n into R^m, with $0 < n < m$, and let M be the image of F. Then the n-coordinate pair (F^{-1}, M) defines a C^r n-subatlas on M.

For further definitions, let M be a fixed C^r n-manifold. An *open set in M* is a subset A of M such that $\phi(A \cap U)$ is open in R^n for every n-coordinate pair (ϕ, U). The reader can verify that M becomes a topological space with this definition of the open sets. If p in M, then a *neighborhood of p* is any open set containing p. Notice M need not be Hausdorff. The concept of Hausdorffness is irrelevant for much of local differential geometry. It becomes relevant in passing from a Riemannian metric to a distance function.

Section 1.2 Smooth Functions

In this section let A be the domain of a function f and assume A is an open subset of the C^r n-manifold M. If f is real valued, then f is C^s *on A* if $f \circ \phi^{-1}$ is C^s on $\phi(A \cap U)$ for every coordinate pair (ϕ, U) on M. Note the independence of r and s. If N is a C^k d-manifold and f is N-valued, and f is C^s *on A* if f is continuous and for every real valued function g, that is C^s on an open domain in N, the composite $g \circ f$ is C^s on $A \cap f^{-1}$ (domain of g). Note the independence of r, k, and s.

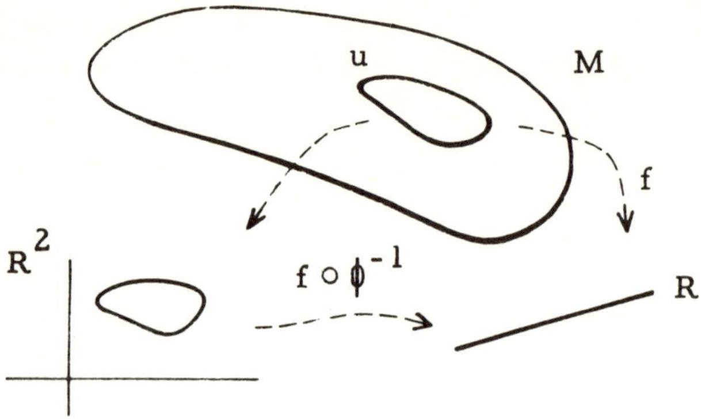

Fig. 1.2 An Induced Map from R^2 into R

The local character of the smoothness of a function is captured in the following definition. Suppose the domain of f is not necessarily open and f is N-valued. If p is in the domain of f, then f is C^s *at* p if there is a neighborhood U of p with f defined on U such that $f|_U$ is C^s on U. As a corollary, if f is C^s at every point in its domain then its domain is open.

Let us now *specialize to* C^∞ *manifolds and* C^∞ *functions*. This is done for convenience chiefly and it allows us to define a tangent vector in a very elegant way. Our concern in these notes is not with "the least possible assumptions" but rather with those concepts that arise naturally in a general situation. The restriction is not too drastic because of the following result due to Whitney: A C^r atlas on a set with $r > 0$ contains a C^∞ atlas (see Munkres[1]). There is an example of Kervaire which exhibits a C^0 atlas on a set which admits no C^1 atlas. For further work on the "equivalence" of differentiable structures see Milnor[1 and 2], Munkres[1 and 2], and Smale[1].

The following list of nine problems are recommended in order to familiarize oneself with the notion of a C^∞ map. In particular the problems are aimed at obtaining numbers 6 and 7 which are often useful. The list (remember A is open in M, which is a C^∞ n-manifold);

1. The map f: $A \to N$ is C^∞ on A iff f is C^∞ at each point p in A.

2. If f: $A \to N$, f is C^∞ on A, and U is an open set contained in A, then $f|_U$ is C^∞ on U.

3. Let U_h be a collection of open sets in M and let f_h: $U_h \to N$ be C^∞ on U_h for each h. If f is a function whose domain is the union of all U_h and if $f|_{U_n} = f_h$ for all h, then f is C^∞ on its domain.

4. If f: $A \to R^k$ is C^∞ on $A \subset R^n$ and g: $B \to R$ is C^∞ on the open set $B \subset R^k$, then $g \circ f$ is C^∞ on $A \cap f^{-1}(B)$.

5. If f : $A \to N$ is C^∞ on $A \subset M$ and (ϕ, U) is a coordinate pair on M, then $f \circ \phi^{-1}$ is C^∞ on $\phi(A \cap U)$.

6. Let P be a C^∞ s-manifold. If F: $A \to N$ is C^∞ on $A \subset M$ and g: $B \to P$ is C^∞ on the open set $B \subset N$, then $g \circ f$ is C^∞ on $A \cap f^{-1}(B)$.

7. The map f: $A \to N$ is C^∞ on $A \subset M$ iff for every coordinate pair (ϕ, U) in a subatlas on N the functions $x_i \circ f$ are C^∞ on $A \cap f^{-1}(U)$, for $i = 1,\ldots,d$ and $x_i = u_i \circ \phi$.

8. If $n \geq k$ and g: $R^n \to R^k$ by $g(a_1,\ldots,a_n) = (a_1,\ldots,a_k)$ then g is C^∞ on R^n. If h: $R^k \to R^n$ by $h(a_1,\ldots,a_k) = (a_1,\ldots,a_k, 0,\ldots,0)$ then h is C^∞ on R^k.

9. Let f and g be real valued functions that are C^∞ on the subsets A and B of M, respectively. Show that $f + g$ and fg are C^∞ on $A \cap B$, where $(f + g)(p) = f(p) + g(p)$ and $(fg)(p) = f(p)g(p)$.

For the record, we can and so do define a Lie group. A *Lie group* G is a group G whose underlying set is also a C^∞ manifold such that the group operations are C^∞, i.e. the map ϕ: $G x G \to G$ where $\phi(g, h) = gh^{-1}$ is C^∞ (see problem 18 and 20).

One last bit of notation, let $C^\infty(A, N)$ denote the set of C^∞ functions mapping an open set A in a manifold M into a manifold N.

Section 1.3 Vectors and vector fields

The definition of a tangent vector generalizes the "directional derivative" in R^n. If X_m is an ordinary (advanced calculus) vector at a point m in R^n and f is a C^∞ function in a neighborhood of m, then define $X_m f = X_m \cdot (\nabla f)_m$, where ∇f is the gradient vector field of f. From the properties of the "dot" product and the operator ∇, it follows that

$$X_m(af + bg) = aX_m f + bX_m g$$

$$X_m(fg) = f(m)X_m g + g(m)X_m f,$$

where g is a C^∞ function in a neighborhood of m and a and b are real numbers. Notice X is *not* normalized to be a unit vector. We generalize now to define a tangent vector on a manifold as an operator on C^∞ functions which obeys the above rules.

Let M be a C^∞ n-manifold. Let m be in M and let $C^\infty(m)$ denote the set of real valued functions that are C^∞ on some neighborhood of m. A *tangent vector at m* is a real valued function X on $C^\infty(m)$ having the following properties:

$$(1) \quad X(f + g) = Xf + Xg, \; X(bf) = b(Xf)$$

$$(2) \quad X(fg) = (Xf)g(m) + f(m)(Xg),$$

where f and g are in $C^\infty(m)$, and b is in R. The set $C^\infty(m)$ is almost a ring (there is a slight problem with domains), and thus a tangent vector is often called a derivation on $C^\infty(m)$.

The *tangent space to M at m*, denoted by M_m, is the set of all tangent vectors at m. It is a vector space over the real field where $(X + Y)f = Xf + Yf$ and $(bX)f = b(Xf)$ for X, Y in M_m, f in $C^\infty(m)$, and b a real number.

Let $x_1,\ldots,x_n$ be a *coordinate system about m* (i. e., m is in the domain of these coordinate functions). We define for each i, a coordinate vector at m, denoted $(\partial/\partial x_i)_m$ by

$$\left(\frac{\partial}{\partial x_i}\right)_m f = \frac{\partial(f \circ \phi^{-1})}{\partial u_i}(\phi(m))$$

where $x_i = u_i \circ \phi$ and the differentiation on the right side is as usual on R^n. The verification of properties (1) and (2) above we leave to the reader. In a moment we show these coordinate vectors form a base for the tangent space at m.

LEMMA. *Let $x_1,\ldots,x_n$ be a coordinate system about m with $x_i(m) = 0$ for all i. Then for every function f in $C^\infty(m)$ there exists n* *functions* $f_1,\ldots,f_n$ *in* $C^\infty(m)$ *with* $f_i(m) = (\partial/\partial x_i)_m f$ *and* $f = f(m) + \Sigma_i x_i f_i$ *in a neighborhood of m. (Note the equality in question is an equality between functions, and f(m) represents a constant function with value f(m); the sum is taken for* $i = 1, 2,\ldots,n$, *and in the future this relevant range is to be understood.)*

Proof. Let ϕ be the coordinate map belonging to the x_i. Let $F = f \circ \phi^{-1}$, and we know F is defined in a ball about the origin in R^n, i.e., in a set $B = [p \text{ in } R^n\text{: distance from origin to } p < r]$. For $(a_1,\ldots,a_n)$ in B we have,

$$F(a_1,\ldots,a_n) = F(a_1,\ldots,a_n) - F(a_1,\ldots,a_{n-1}, 0) +$$

$$F(a_1,\ldots,a_{n-1}, 0) - F(a_1,\ldots,a_{n-2}, 0, 0) +\ldots+$$

$$F(a_1, 0,\ldots,0) - F(0,\ldots,0) + F(0,\ldots,0) =$$

$$= \Sigma_i F(a_1,\ldots,a_{i-1}, ta_i, 0,\ldots,0)]_0^1 + F(0,\ldots 0)$$

$$= F(0,\ldots,0) + \Sigma_i \int_0^1 \frac{\partial F}{\partial u_i}(a_1,\ldots,a_{n-1}, ta_i, 0,\ldots,0)a_i dt$$

$$= F(0,\ldots,0) + \Sigma_i a_i F_i(a_1,\ldots,a_n), \text{ where}$$

$$F_i(a_1,\ldots,a_n) = \int_0^1 \frac{\partial F}{\partial u_i}(a_1,\ldots,a_{i-1}, ta_i, 0,\ldots,0)dt$$

is C^∞ in B since $(\partial F/\partial u_i)$ is C^∞. Let $f_i = F_i \circ \phi$ and the lemma is proved. //

THEOREM. *Let M be a* C^∞ *n-manifold and let* $x_1,\ldots,x_n$ *be a coordinate system about m in M. Then if X in* M_m, $X = \Sigma_i (Xx_i)(\partial/\partial x_i)_m$, *and the coordinate vectors form a base for* M_m *which thus has dimension n.*

Proof. We first prove the stated representation. Take X in M_m and f in $C^\infty(m)$. If $x_i(m) \neq 0$ for all i, let $y_i = x_i - x_i(m)$. Then apply the lemma to f with respect to the coordinate system $y_i,\ldots,y_n$ and notice $(\partial f/\partial y_i)(m) = (\partial f/\partial x_i)(m)$. Next we see if c a constant map then

$$X(c) = cX(1) = c(1X(1) + 1X(1)) = 2cX(1)$$

which implies $cX(1) = 0$ and $X(c) = 0$. Thus $Xf = X(f(m) + \Sigma_i y_i f_i)$

$$= \Sigma_i[(Xy_i)f_i(m) + y_i(m)(Xf_i)]$$

$$= \Sigma_i X(x_i - x_i(m))f_i(m)$$

$$= \Sigma_i (Xx_i)(\partial f/\partial x_i)(m)$$

which proves the required representation. If $Y = \Sigma_i a_i(\partial/\partial x_i) = 0$ then $0 = Yx_j = a_j$, thus the coordinate vectors are independent and span M_m.//

A vector *field* X on a set A is a mapping that assigns to each point p in A a vector X_p in M_p. A field X is C^∞ on A if A is open and for each real valued function f that is C^∞ on B, the function $(Xf)(p) = X_p f$ is C^∞ on $A \cap B$. If X and Y are C^∞ vector fields on A their *bracket* is a C^∞ vector field $[X, Y]$ on A defined by $[X, Y]_p f = X_p(Yf) - Y_p(Xf)$.

If f and g are C^∞ functions, it is trivial that $[X, Y](f + g) = [X, Y]f + [X, Y]g$, and $[X, Y](af) = a[X, Y]f$ for a in R. To check the product property, consider

$$\begin{aligned}[X, Y](fg) &= X(Y(fg)) - Y(X(fg)) \\ &= X(fYg + gYf) - Y(fXg + gXf) \\ &= fXYg + (Xf)(Yg) + (Xg)(Yf) + gXYf \\ &\quad - fYXg - (Yf)(Xg) - (Yg)(Xf) - gYXf \\ &= f[X, Y]g + g[X, Y]f.\end{aligned}$$

Thus $[X, Y]$ is a vector field and the proof of its C^∞ nature we leave as a problem.

For later use, notice that $[X, Y] = -[Y, X]$, $[X, X] = 0$, and the bracket is linear in each slot with respect to addition, i.e., $[X_1 + X_2, Y] = [X_1, Y] + [X_2, Y]$. However, $[fX, gY] = f(Xg)Y - g(Yf)X + fg[X, Y]$, and it is this property that prevents the bracket mapping from being a tensor (problem 10). Problem 13 gives a geometric interpretation of the bracket, and in section 9.1 there are applications involving integrability conditions. For example, if $x_1,\ldots,x_n$ is a coordinate system then $[\partial/\partial x_i, \partial/\partial x_j] = 0$ for all i and j (since cross partial derivatives of C^∞ functions are equal), and actually this condition on n independent vector fields is sufficient to imply the fields are coordinate vector fields (section 9.1).

The bracket operation also satisfies the following expression which is called the *Jacobi identity,*

$$[X, [Y, Z]] + [Z, [X, Y]] + [Y, [Z, X]] = 0$$

where X, Y, and Z are C^∞ fields with a common domain.

Section 1.4 The Jacobian of a map

Let M and N be C^∞ manifolds of dimensions n and k respectively. We defined above the concept of a C^∞ map f from M into N. Such a map induces a linear transformation from each tangent space M_m into the tangent space $N_{f(m)}$. This linear map is called the *Jacobian map* or the *differential of f* and we denote it by f_* (often it is denoted df, but we reserve the symbol d for the exterior derivative operator). Let X be in M_m and we define f_*X as a vector at $f(m)$ in N by taking a function g which is C^∞ in a neighborhood of $f(m)$ and setting $(f_*X)g = X(g \circ f)$. It is trivial to check that f_*X is a vector at $f(m)$ and the map f_* is linear.

By selecting a coordinate system $x_1,\ldots,x_n$ about m and another $y_1,\ldots,y_k$ about $f(m)$, we can determine a matrix representation for f_* which is called the *Jacobian matrix* of f_* with respect to the chosen coordinate systems. Let $X_i = \partial/\partial x_i$, $Y_j = \partial/\partial y_j$, thus $X_1,\ldots,X_n$, at m, form a base for M_m and we compute f_* by computing its action on this base. Namely, $f_*X_i = \Sigma_j(f_*X_i)y_jY_j$ by the representation theorem above, hence the matrix in question is the matrix $((f_*X_i)y_j) = (\partial(y_j \circ f)/\partial x_i)$ for $1 \leq i \leq n$ and $1 \leq j \leq k$.

The implicit function theorem and the inverse function theorem can be applied and formulated in this language. The former we postpone, since we do not really need it for some time (see problem 16) but the latter is both useful and instructive. First a definition. A *diffeomorphism* is a map $f\colon\ M \to N$ that is 1:1 and onto with both f and $f^{-1}C^\infty$, and if such an f exists, then M is *diffeomorphic* to N.

THEOREM. (*Inverse function*) *Let M and N be C^∞ n-manifolds and let $f\colon M \to N$ be C^∞. If for m in M, the Jacobian f_* at m is an isomorphism of M_m onto $N_{f(m)}$ then there is a neighborhood U of m and a neighborhood V of $f(m)$ such that f is a diffeomorphism from U to V (i.e., f is a local diffeomorphism about m).*

We leave it to the reader to choose a coordinate system on both sides and apply the theorem from advanced calculus to obtain the result. Notice the C^∞ demand of f *and* f^{-1} implies the theorem could be stated as a *necessary* as well as a sufficient condition for the existence of a local inverse. If one only demands continuity of the inverse, then the map $x \to x^3$ provides a homeomorphism of R onto R whose Jacobian is singular at the origin.

Now consider the behavior of the Jacobian with respect to composite maps. Let g be a C^∞ map of N into the C^∞ manifold L. Then at each m in M, $(g \circ f)_* = g_* \circ f_*$, for if h is a C^∞ function about $g(f(m))$ and X in M_m then $((g \circ f)_*X)h = X(h \circ g \circ f) = (f_*X)(h \circ g) = (g_*(f_*X))h$. In terms of coordinate systems, the above computation exhibits the chain rule and a multiplicative behavior of Jacobian matrices. When f is a diffeomorphism of M into N, and X and Y are C^∞ fields on M, then f_*X and f_*Y are C^∞ fields on N with $f_*[X, Y] = [f_*X, f_*Y]$.

Section 1.5 Curves and integral curves

In these notes curves will be viewed as a special case of mappings, thus we will deal with "parameterized curves" almost exclusively. A *curve* in M is a C^∞ map σ from an open subset of R into M. Often we speak of a curve σ from $[a, b]$ into M where $[a, b]$ is a closed interval of real numbers, and in this case it is assumed the domain of σ is actually an open set in R containing $[a, b]$.

Let σ be a curve in M with domain U. For each t in U define the *tangent of σ at t* to be the vector $T(t)$, or $T_\sigma(t)$, at $\sigma(t)$ where $T(t) = \sigma_*(d/dt)_t$ and d/dt denotes the usual differentiation operator of real valued C^∞ functions on R. Thus if $x_1,\ldots,x_n$ a coordinate system about $\sigma(t)$, then $T(t) = \Sigma_i(d(x_i \circ \sigma)/dt)_t(\partial/\partial x_i)_{\sigma(t)}$. By differentiating the coordinate parameter functions $x_i \circ \sigma(t)$ one determines the coefficients of $T(t)$ with respect to the coordinate vectors associated with the co-

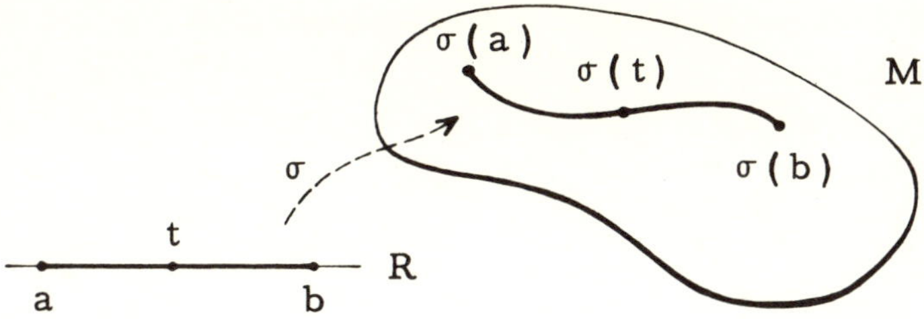

Fig. 1.3 A Curve

ordinate system. Notice this $T(t)$ is the usual "velocity" vector associated with a parameterized curve in R^3.

Having the idea of curve and tangent vector we can give a geometric description of the Jacobian f_* associated with the map f: $M \to N$. For X in M_m choose any curve σ on M with $\sigma(0) = m$ and $T_\sigma(0) = X$. Then $f \circ \sigma$ is a curve on N with $f \circ \sigma(0) = f(m)$ and indeed $f_*X = T_{f\circ\sigma}(0)$. Thus we "fill in the vector by a curve, map the curve to N, and take the new tangent vector." This device is very useful if one knows geometrically the behavior of certain curves; e.g., let $M = [(x, y, z)$ in R^3: $x^2 + y^2 = 1]$, let S be the unit sphere in R^3, and let f: $M \to S$ by $f(x, y, z) = (x, y, 0)$. The particular f just defined is called the "sphere map" or the "Gauss map" from M to S, since it essentially uses a unit normal vector field to M in its definition. Its Jacobian should be trivial to compute at each point from the above remarks.

We carry the idea of "filling in a vector" to a classical setting. Let X be a C^∞ vector field on the manifold M. A curve σ is an *integral curve* of X if whenever $\sigma(t)$ is in the domain of X then $T_\sigma(t) = X_{\sigma(t)}$. Thus we say the curve σ "fits" X, and suggest the physical example of the velocity vector field (which gives X) of a steady fluid flow and its streamlines (which give integral curves). The local existence of integral curves is guaranteed by the theory of ordinary differential equations.

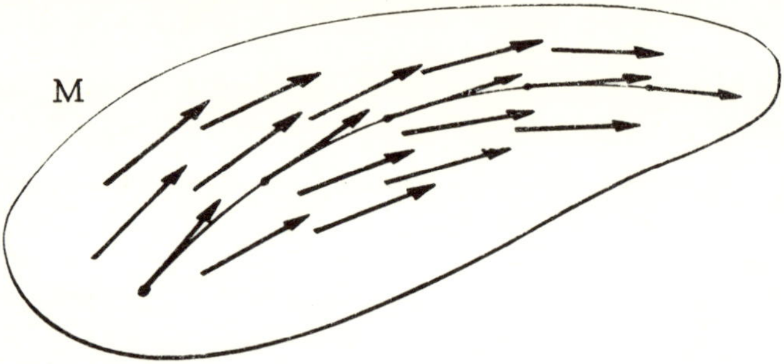

Fig. 1.4 An Integral Curve of a Vector Field

THEOREM. *Let X be a C^∞ vector field on M and let m be a point in the domain of X. Then for any real number b there exists a real number $r > 0$ and a unique curve σ: $(b - r, b + r) \to M$ such that $\sigma(b) = m$ and σ an integral curve of X.*

Proof. Let $x_1,\ldots,x_n$ be a coordinate system about m whose domain U is contained in the domain of X. Let $X = \Sigma_i f_i(\partial/\partial x_i)$ define C^∞ real valued functions f_i on U. Then the condition that a curve σ be an integral curve of X becomes the condition

$$\frac{d(x_i \circ \sigma)}{dt} = f_i \circ \sigma$$

on the domain of σ, or writing (improperly) as usual $x_i(t) = x_i \circ \sigma(t)$, we have the system of first order ordinary differential equations

$$\frac{dx_i}{dt} = f_i(x_1,\ldots,x_n),$$

for $i = 1,\ldots,n$. Apply an existence and uniqueness theorem from differential equation theory to obtain $r > 0$ and functions $x_i(t)$ that define σ on the specified range with the required properties.//

Actually the theorem from differential equations gives much more than the above conclusion for it includes the C^∞ dependence of solutions as we vary the initial parameter b and the point m (see section 9.3). We return to this later when discussing the existence of geodesics and the exponential map (sections 5.1 and 9.3). For global ramifications see Palais[2] or Lang.

It is convenient to define a *broken* C^∞ *curve* σ *on an interval* $[a, b]$ to be a continuous map σ from $[a, b]$ into M which is C^∞ on each of a finite number of subintervals $[a, b_1]$, $[b_1. b_2],\ldots,[b_{k-1}, b]$.

Section 1.6 Submanifolds

A C^∞ k-manifold M is a *submanifold* of a C^∞ n-manifold $\overline{M}$ if for every point p in M there is a coordinate neighborhood $\overline{U}$ of $\overline{M}$ with coordinate functions $\overline{x}_1,\ldots,\overline{x}_n$ such that the set $U = [m \text{ in } \overline{U}\colon\ \overline{x}_{k+1}(m) = \ldots = \overline{x}_n(m) = 0]$ is a coordinate neighborhood of p in M with coordinate functions $x_1 = \overline{x}_1|_U,\ldots,x_k = \overline{x}_k|_U$. These coordinate systems are called *special* or *adapted* coordinate systems.

Notice it is not required that $M \cap \overline{U} = U$ so "slices" of M may approach other "slices" of M in $\overline{M}$ (see problem 17), and hence the topology on M may not be the relative topology. The definition of submanifold implies M is a subset of $\overline{M}$ and $k \leq n$. Letting $i\colon\ M \to \overline{M}$ be the inclusion map, then i is C^∞ since $\overline{x}_j \circ i$ are C^∞ maps for all special coordinate functions. The inclusion map is also an imbedding (see below) since the Jacobian i_* is non-singular, i.e., $i_*(\partial/\partial x_j(p) = \partial/\partial\overline{x}_j(p)$ for $j = 1,\ldots,k$. In these notes we will identify a tangent vector X in M_p with its image in $\overline{M}_p$ unless there is a possibility of confusion (just as we identify p and $i(p)$).

To make some more standard definitions, let M and $\overline{M}$ be C^∞ manifolds and let f be a C^∞ map of M into $\overline{M}$. If f_* is non-singular (thus f_* has no kernel) at each point p of M, then f is called an *immersion* of M into $\overline{M}$. If in addition, f is univalent, then f is called an *imbedding* of M into $\overline{M}$. A subset M' of $\overline{M}$ is called an *immersed submanifold* if there exists a manifold M and an immersion $f\colon\ M \to \overline{M}$ such that $f(M) = M'$. (Thus an immersion is a "local imbedding with self-intersections.") One can verify (problem 17) that if $f\colon M \to \overline{M}$ is an imbedding and $M' = f(M)$, then by defining a differentiable structure on M' so f becomes a diffeomorphism, M' becomes a submanifold of $\overline{M}$ (see Helgason, p. 23).

For examples of submanifolds see the examples 5, 6, and 7 at the end of section 1.1.

It is convenient to define a *base field* on a set A contained in an n-manifold to be a set of n vector fields that are independent at each point of A. When each field in a base field is C^∞, then the base field is C^∞. Since a set of coordinate fields is a C^∞ base field on the coordinate domain, we know C^∞ base fields always exist locally. A C^∞ base field does *not* necessarily exist over a whole manifold (consider the 2-sphere, S^2); indeed, the manifold is called *parallelizable* if it admits a global C^∞ base field.

We now define a concept which we will often use. Let M be a submanifold of $\overline{M}$ as described above. An *$\overline{M}$-vector field* Z that is C^∞ *on* M (or C^∞ on an open set A in M) is a map that assigns to each p in M (or p in A) a vector Z_p in $\overline{M}_p$ such that if $X_1,\ldots,X_n$ is any C^∞ base field on a neighborhood $\overline{U}$ of p and $Z_m = \Sigma_i^n a_i(m)(X_i)_m$ for m in $M \cap \overline{U}$ then the real valued functions a_i are C^∞ on $M \cap U$ for all i. Notice Z_p is *not* necessarily tangent to M. Since the restriction to M, of a C^∞ function on $\overline{M}$, is a C^∞ function on M, it follows if Z is C^∞ on $\overline{M}$ then $Z|_M$ is an $\overline{M}$-vector field that is C^∞ on M.

Problems (For problems 1 thru 9 see pages 4 and 5)

10. Let $W_1,\ldots,W_n$ be a C^∞ base field on an open set U in a manifold M and let $X = \Sigma_{i=1}^n f_i W_i$ be a vector field on U. Show X is C^∞ on U iff the functions f_i are C^∞ on U for all i. If Y and Z are C^∞ fields on U show $[Y, Z]$ is C^∞. Show that a coordinate field $\partial/\partial x_i$ is C^∞ on its domain. If X_p is a given vector at p in M show there is a C^∞ field $\overline{X}$ on a neighborhood of p with $\overline{X}_p = X_p$. If $x_1,\ldots,x_n$ is a coordinate system with domain U and $A = \Sigma a_i(\partial/\partial x_i)$ and $B = \Sigma b_j(\partial/\partial x_j)$ are C^∞ fields on U then find the representation of $[A, B]$ in terms of the coordinate vector fields. Show $[fX, gY] = f(Xg)Y - g(Yf)X + fg[X, Y]$ where X and Y are C^∞ fields on U and f and g are in $C^\infty(U, R)$. Prove the Jacoby identity.

11. Let A, B and C be in $C^\infty(R^3, R)$ with $B \neq 0$ anywhere. Let $V = Ai + Bj + Ck$, $X = -Bi + Aj$, and $Y = -Cj + Bk$ (advanced calculus notation). For p in R^3, let $P_p = [Z \text{ in } (R^3)_p\colon Z \cdot V_p = 0]$. Show P_p is a two-dimensional space of vectors at each point by showing X_p and Y_p are a base for P_p. Show $[X, Y]_p$ lies in P_p iff $V_p \cdot (\text{curl } V)_p = 0$. If there is a function f in $C^\infty(R^3, R)$ with grad $f \neq 0$ such that P_p is the tangent plane

to the constant surface of f thru p show $V_p \cdot (\text{curl } V)_p = 0$ (see section 9.1).

Instead of seeking surfaces that are orthogonal to V (as above), one could seek surfaces whose tangent plane contains V and then one has a "geometric quasi-linear partial differential equation of the first order." Integral curves of V are called *characteristics* of the "equation." One generates solution surfaces by taking a non-characteristic curve (an "initial value" curve) and considering the surface formed by characteristics thru the initial value curve. Show two solution surfaces must intersect along a characteristic. Show there are an infinite number of solution surfaces thru one characteristic. Can there there be an initial value curve with no solution thru it?

12. Let f: $R^2 \to R^2$ by $f(a, b) = (a^2 - 2b, 4a^3b^2)$ and let g: $R^2 \to R^3$ by $g(u, v) = (u^2v + v^2, u - 2v^3, ve^u)$. Compute a matrix for f_* at (1, 2) and g_* at any (u, v). Find $g_*(4\partial/\partial x - \partial/\partial y)_{(0,1)}$. Find integral curves for the vector field $X = yi + yj + 2k$ on R^3. Find a coordinate system x_1, x_2, x_3 on R^3 such that $\partial/\partial x_1 = 2i + 3j - k$ at all points.

13. Let X and Y be C^∞ fields about m in M. For small $t \geq 0$ define the curve $\sigma(t)$ as follows: go t parameter units on X integral curve thru m to p_1, go t units on Y curve thru p_1 to p_2, go t units on $(-X)$ curve thru p_2 to p_3, go t units on $(-Y)$ curve thru p_3 to $\sigma(t)$. If $\gamma(t) = \sigma(\sqrt{t})$ show $T_\gamma(0) = [X, Y]_m$. (Hint: use the lemma in section 9.1 and partial Taylor series.)

14. Let M and N be manifolds with M connected and let f and g be C^∞ maps of M into N. Show $f_* \equiv 0$ iff f is a constant map. If $f(m) = g(m)$ at one m in M and $f_* \equiv g_*$ at all points show $f = g$.

15. Let f be in $C^\infty(M, R)$ and define the *differential of f, df,* to be the linear map of M_m into R where $(df)_m(X_m) = X_mf$. Show $f_*(X_m) = [(df)_m(X)](\partial/\partial t)$ where t is the identity coordinate function on R. It is because of this case that in a general case the Jacobian f_* is often called the "differential of f".

16. Prove the Inverse Function Theorem (p. 10). State and prove a version of the Implicit Function Theorem of advanced calculus in terms of the Jacobian map.

17. Prove the last sentence in the third paragraph of section 1.6. Show that the image of a regular ($\sigma_* \neq 0$) univalent curve σ mapping an open interval into a manifold M is a one-dimensional submanifold of M. Let X be a unit constant vector field on R^2 with irrational slope. Let T be the set of equivalence classes on R^2 where $(a, b) \sim (c. d)$ iff $a - c = n$ and $(b - d) = m$ for integers m and n. Show T is a two-dimensional manifold (which is called the *flat torus*) in a natural way. Show X induces a vector field on T such that the image of one integral curve of X defines a one-dimensional submanifold of T that is dense in T.

18. Let M_1 and M_2 be C^∞ manifolds. Let π_i: $M_1 \times M_2 \to M_i$ by $\pi_i(m_1, m_2) = m_i$ for $i = 1, 2$. Define a C^∞ structure on $M_1 \times M_2$ so π_i are C^∞. Show $(M_1 \times M_2)_{(m_1, m_2)}$ is naturally isomorphic to $(M_1)_{m_1} \times (M_2)_{m_2}$.

19. Let M be a C^∞ n-manifold. Let $T(M) = [(m, X)\text{: } X \text{ in } M_m]$, and let π: $T(M) \to M$ by $\pi(m, X) = m$. If (ϕ, U) is a coordinate pair on M with $x_i = u_i \circ \phi$ let $\bar{U} = \pi^{-1}(U)$, $\bar{x}_i = x_i \circ \pi$, and for (m, X) in $\bar{U}$ let $\dot{x}_i(m, X) = a_i$ if $X = \Sigma a_i(\partial/\partial x_i)$. Let $\bar{\phi}$: $\bar{U} \to R^{2n}$ so $u_i \circ \bar{\phi} = \bar{x}_i$ and $u_{n+i} \circ \bar{\phi} = \dot{x}_i$ for $i = 1,\ldots,n$. Show the subatlas of pairs $(\bar{\phi}, \bar{U})$ defines a C^∞ structure on $T(M)$ which is called the *tangent bundle of M*. If f is a C^∞ map of M into N show f_* induces a C^∞ map of $T(M)$ into $T(N)$.

20. Let G be a Lie group. If g in G let L_g, R_g, and A_g denote the maps of G into G defined by $L_g(h) = gh$, $R_g(h) = hg$, and $A_g(h) = ghg^{-1}$. Show L_g, R_g, and A_g are C^∞. A vector field X on G is *left invariant* if $(L_g)_* X_g = X_{gh}$ for all g and h. Show a left invariant field is C^∞ and is completely determined by its value at the identity e. If X and Y are left invariant, show $[X, Y]$ is left invariant. The set of left invariant vector fields on G forms an n dimensional vector space called the *Lie algebra of* G which is denoted by $\mathfrak{g}$. Define a *one-parameter* subgroup of G to be the image of a C^∞ homormorphism of R into G. Show there is a 1:1 correspondence between one-parameter subgroups and integral curves of left invariant vector fields thru e. Show the map $(g, h) \to gh^{-1}$ is C^∞ from $G \times G$ into G

iff the maps $(g, h) \to gh$ and $g \to g^{-1}$ are C^∞.

21. Let $G = GL(n, R)$ and for a matrix g in G let $u_{ij}(g) = g_{ij}$ (see example 3). Call u_{ij} the natural coordinate functions on G. Write $u_{ij} \cdot L_g$ as a linear combination of the natural coordinate functions. Let X_{ij} be the unique left invariant field on G with $X_{ij}(e) = (\partial/\partial u_{ij})(e)$ where e is the identity element. Compute X_{ij} as a field on G in terms of the coordinate vector fields. Compute $[X_{ij}, X_{rs}]$. If $A(t)$ is a C^∞ curve in G with $A(0) = e$ and $A(t)$ orthogonal for all t show $dA/dt = (da_{ij}/dt)$ is a skew-symmetric matrix for $t = 0$.

22. Let M be a C^∞ n-manifold. Let $B(M) = [(m;\ e_1,\ldots,e_n)\colon\ m$ in M and $e_1,\ldots,e_n$ an ordered based of $M_m]$. Let $\pi\colon\ B(M) \to M$ by $\pi(m;\ e_1,\ldots,e_n) = m$. If (ϕ, U) a coordinate pair on M with $x_i = u_i \cdot \phi$, let $(\bar{\phi}, \bar{U})$ be a coordinate pair on $B(M)$ with $\bar{U} = \pi^{-1}(U)$ and $\bar{\phi}\colon\ \bar{U} \to R^{n+n^2}$ by the coordinate functions $\bar{x}_1,\ldots,\bar{x}_n$, $x_{11}, x_{12},\ldots,x_{nn}$ where $\bar{x}_i = x_i \cdot \pi$ and if $b = (m;\ e_1,\ldots,e_n)$ then $e_j = \sum_{i=1}^n x_{ij}(b)(\partial/\partial x_i)$. Show the subatlas of pairs $(\bar{\phi}, \bar{U})$ defines a C^∞ structure on $B(M)$ which is called the *bundle of bases over M*. For g in $GL(n, R)$ let $R_g\colon\ B(M) \to B(M)$ by $R_g(b) \equiv bg \equiv (m;\ \sum_{i=1}^n g_{i1}e_i,\ \sum_i g_{i2}e_i,\ldots,\sum_i g_{in}e_i)$ if $b = (m;\ e_1,\ldots e_n)$. Show R_g is C^∞. Let $s_U\colon\ U \to B(M)$ by $s_U(m) = (m;\ (\partial/\partial x_1)_m, \ldots,(\partial/\partial x_n)_m)$ for m in U. Show s_U is C^∞ and $\pi \cdot s_U$ is the identity on U. The map s_U is called the *coordinate section map* over U. Let $\hat{\phi}\colon\ U \times GL(n, R) \to \bar{U}$ by $\hat{\phi}(m, g) = R_g \cdot s_U(m) = s_U(m)g$. Show $\hat{\phi}$ is a diffeo onto its image. The map $\hat{\phi}$ is called a *strip map*. If (ϕ, U) and (ψ, V) are coordinate pairs on M define $s_{UV}\colon\ U \cap V \to GL(n, R)$ by $s_{UV}(m) = g$ if $s_U(m)g = s_V(m)$. Show s_{UV} is C^∞; it is called a *structural function* for $B(M)$. Show $(bg_1)g_2 = b(g_1g_2)$ which justifies the name *right action* for R_g. For fixed b in $B(M)$ let $f_b\colon\ GL(n, R) \to B(M)$ by $f_b(g) = bg$. Show f_b is C^∞. Call the set $F_m = \pi^{-1}(m)$ the (vertical) *fiber over m* in M. Show F_m is an n^2 – submanifold of $B(M)$ and f_b is a diffo of $GL(n, R)$ onto $F_{\pi(b)}$. If $\pi(b) = \pi(c)$, show $f_c^{-1} \circ f_b$ is a left translation on $GL(n, R)$. A vector X on $B(M)$ such that $\pi_*(X) = 0$ is called a *vertical* vector. For b in $B(M)$, let $E_{ij}(b) = (f_b)_* X_{ij}(e)$ define a vector $E_{ij}(b)$ (see problem 21). Show E_{ij} is a global C^∞ vertical vector field on $B(M)$. Compute $[E_{ij}, E_{rs}]$.

2. Hypersurfaces of R^n

In a very real sense, this chapter and the next are too special, i.e.,, much of the theory belongs to an arbitrary submanifold of a "semi-Riemannian" manifold. We specialize because we can obtain many of the concepts and results of classical differential geometry quickly and easily. In so doing, we hope to develop the "geometric" intuition of the reader sufficiently to make later generalizations and definitions seem natural.

Section 2.1 The standard connexion on R^n.

Recall in section 1.3 we shifted the classical notion of a vector from a " directed line segment" to an operator on functions, i.e., if $X = a\vec{i} + b\vec{j} + c\vec{k}$ is a familiar vector on R^3 from advanced calculus, then we rewrite $X = a(\partial/\partial x) + b(\partial/\partial y) + c(\partial/\partial z)$ so if f is a real valued C^∞ function on R^3, then Xf is a *derivative of f in the direction X,*

$$Xf = X \cdot \nabla f = a\frac{\partial f}{\partial x} + b\frac{\partial f}{\partial y} + c\frac{\partial f}{\partial z}.$$

Notice that X need not be a unit vector. When a, b, and c are C^∞ functions on R^3 themselves (possibly constant functions), then X is a C^∞ field and Xf is a C^∞ real valued function on R^3,

$$(Xf)(p) = X_p f = a(p)\frac{\partial f}{\partial x}(p) + b(p)\frac{\partial f}{\partial y}(p) + c(p)\frac{\partial f}{\partial z}(p).$$

Since both of the representations of a vector field X given above are awkward to write, let us simply write $X = (a, b, c)$, thus giving X by giving the coefficient functions (or constants) a, b, and c of the global base field $\partial/\partial x$, $\partial/\partial y$, $\partial/\partial z$ on R^3.

We now define the *derivative of a vector field Y in a direction X.* Let X be a vector at p in R^n and let $Y = (y_1,\ldots,y_n)$ be a C^∞ field about p, thus each y_i is a C^∞ real valued function on the domain of Y which includes p. The *covariant derivative of Y in the direction X* is the vector $\bar{D}_X Y = (X_p y_1,\ldots,X_p y_n)$ as a vector at p. If X and Y are C^∞ fields with the same domain A, then $\bar{D}_X Y$ is a C^∞ field with domain A.

For example take R^3, let $X = (a, b, c)$, let $Y = (xy^2 + 4z, y^2 - x, x + z^3)$. and then

$$\bar{D}_X Y = [X \cdot (y^2, 2xy, 4), X \cdot (-1, 2y, 0), X \cdot (1, 0, 3z^2)]$$

$$= (ay^2 + 2xyb + 4c, -a + 2yb, a + 3z^2c),$$

where a, b, and c may be functions or constants.

The properties of $\bar{D}$ which we now list are one of the main analytic tools of these notes. Let X and W be vectors at p in R^n, let Y and Z be C^∞ fields about p, and let f be a C^∞ real valued function about p. Then

(1) $\bar{D}_X(Y + Z) = \bar{D}_X Y + \bar{D}_X Z$

(2) $\bar{D}_{X+W}(Y) = \bar{D}_X Y + \bar{D}_W Y$

(3) $\bar{D}_{f(p)X} Y = f(p)\bar{D}_X Y$

(4) $\bar{D}_X(fY) = (Xf)Y_p + f(p)\bar{D}_X Y$.

These follow directly from the definition of $\bar{D}$. It is important to notice $\bar{D}_X Y$ can be computed once one knows Y along a curve σ that fits X, i.e., if $\sigma(0) = p$ and $T_\sigma(0) = T(0) = X_p$. For let $Y_{\sigma(t)} = (y_1(t), \ldots, y_n(t))$ and then $\bar{D}_X Y = (\frac{dy_1}{dt}(0), \ldots, \frac{dy_n}{dt}(0))$ since by the chain rule, $\frac{dy_i}{dt}(0) = \sum_{j=1}^{n} \frac{\partial y_i}{\partial u_j}(p) \frac{du_j}{dt}(0) = X_p \cdot (\nabla y_i)_p$ and $T(0) = X_p$. Thus if Y is an R^n – vector field that is C^∞ on the curve σ with tangent T, then $\bar{D}_T Y$ is a well-defined R^n-vector field that is C^∞ on σ.

Using the operator $\bar{D}$, we can define parallel vector fields along a curve and geodesics. Let σ be a C^∞ curve (in R^n) with tangent T and let Y be an R^n-vector field that is C^∞ on σ. The field Y is *parallel along* σ if $\bar{D}_T Y \equiv 0$ along σ. The curve σ is a *geodesic* if $\bar{D}_T T \equiv 0$, i.e. if its tangent T is parallel along σ.

It is trivial to see these are the usual concepts of parallel fields and geodesics in R^n; for let $\sigma(t) = (a_1(t), \ldots, a_n(t))$ and $Y_{\sigma(t)} = (y_1(t), \ldots, y_n(t))$. Then $\bar{D}_T Y = (dy_1/dt, \ldots, dy_n/dt) = 0$ iff each $y_i(t)$ is a constant function of t, so Y is a "constant" vector field of R^n evaluated on σ. The curve σ is a geodesic iff $\bar{D}_T T = (d^2a_1/dt^2,\ldots,d^2a_n/dt^2) = 0$, and this implies $a_i(t) = c_i t + d_i$ are linear functions of t so σ is a linear parameterization of a straight line.

Notice that the parameterization of a curve is important in the definition of a geodesic.

The generalization of the definition of *covariant differentiation* or a *connexion* on any C^∞ manifold M is clear, i.e. we merely demand the existence of an operator D which satisfies the above four properties (listed for $\bar{D}$) and assigns to C^∞ vector fields X and Y with the domain A, a C^∞ field $D_X Y$ on A. Notice there can be more than one connexion on a manifold. In the case of "semi-Riemannian" manifolds however there exists one connexion which fits the "semi-Riemannian" structure nicely, and in the case of R^n, $\bar{D}$ is this nice connexion. In fact, we now explain how $\bar{D}$ is "nice."

Henceforth, denote the usual dot product or inner product of vectors Y and Z tangent to R^n by $\langle Y, Z\rangle$. Thus if $Y = (y_1,\ldots,y_n)$ and $Z = z_1,\ldots,z_n)$, then $\langle Y, Z\rangle = \sum_{i=1}^n y_i z_i$. If Y and Z are C^∞ fields with domain A, then $\langle Y, Z\rangle$ is a C^∞ function with domain A. One checks easily that

(5) $\quad \bar{D}_Y Z - \bar{D}_Z Y = [Y, Z]$ on A, and

(6) $\quad X_p\langle Y,Z\rangle = \langle \bar{D}_X Y, Z\rangle_p + \langle Y, \bar{D}_X Z\rangle_p$

for any vector X at p in A.

We now generalize and fix some terminology. A *Riemannian manifold* is a C^∞ manifold M on which one has singled out a C^∞ real valued, bilinear, symmetric, and positive definite function $\langle\, ,\, \rangle$ on ordered pairs of tangent vectors at each point. Thus if X, Y and Z are in M_p, then X, Y is a real number and $\langle\, ,\rangle$ satisfies the following properties:

(a) (symmetric) $\langle X, Y\rangle = \langle Y, X\rangle$,

(b) (bilinear) $\langle X + Y, Z\rangle = \langle X, Z\rangle + \langle Y, Z\rangle$
$\langle aX, Y\rangle = a\langle X, Y\rangle$ for a in R,

(c) (positive definite) $\langle X, X\rangle > 0$ for all $X \neq 0$,

(d) (C^∞) if X and Y are C^∞ fields with domain A then

$$\langle X, Y\rangle_p = \langle X_p, Y_p\rangle \text{ is a } C^\infty \text{ function on } A.$$

When (c) is replaced by

(c') (non-singular) $\langle X, Y\rangle = 0$ for all X implies $Y = 0$,

then M is a *semi-Riemannian* (or pseudo-Riemannian) manifold. In either case, the functional $\langle\ ,\ \rangle$ is called the *inner product, the metric tensor*, the *Riemannian metric*, or the *infinitesimal metric* of M. Notice the word "metric" in the preceding sentence is not referring to a metric function (distance function) in the topological sense. In Chapter 6, the connexion of the concepts is clarified. It is also customary to require a semi-Riemannian manifold to be Hausdorff; however, as far as the local differential geometry is concerned, this is irrelevant so the restriction is not enforced at this time.

If D is a C^∞ connexion in a semi-Riemannian manifold M, then D is a *Riemannian connexion* if it satisfies the above properties (5) and (6). In Chapter 6, the existence of Riemannian manifolds is discussed and the fundamental theorem asserting the existence and uniqueness of a Riemannian connexion is proved. In section 2.3 one sees that many hypersurfaces in $R^n (n \geq 3)$ provide examples of Riemannian manifolds with a Riemannian connexion.

Section 2.2 The sphere map and the Weingarten map.

An $(n-1)$-submanifold of an n-manifold is called a *hypersurface*. Throughout this section let M be a hypersurface of R^n, let $\bar{D}$ be the natural connexion on R^n, and assume N is a unit normal vector field that is C^∞ on M. Thus $\langle N_p, N_p\rangle = 1$ and $\langle N_p, X\rangle = 0$ for all p in M and X in M_p. Such an N always exists locally.

For any p in M and any vector X in M_p, define the linear map $L\colon M_p \to M_p$ by

$$(7) \qquad L(X) = \bar{D}_X N.$$

The vector $L(X)$ lies in M_p, since $0 = X\langle N, N\rangle = 2\langle L(X), N\rangle$ by property (6) for $\bar{D}$. The map L is linear by properties (2) and (3). The map L is called the *Weingarten map*, and in the case of R^n, it has a geometric interpretation as the Jacobian of the sphere map (Gauss map) which we now explain.

Let $N = (a_1,\ldots,a_n)$, so the a_i are real valued C^∞ functions on M and $\Sigma(a)^2 = 1$. Then the map η: $M \to S^{n-1}$ defined by $\eta(p) = (a_1(p), \ldots, a_n(p))$ in R^n, is a C^∞ map of M into the unit $(n-1)$-sphere S^{n-1}, and η is called the *sphere map* (or *Gauss map*). If X in M_p and $\sigma(t)$ is a curve fitting X (so $\sigma(0) = p$ and $T_\sigma(0) = X$), then $\eta \circ \sigma(t) = (a_1 \circ \sigma(t),\ldots,a_n \circ \sigma(t))$ and

$$\eta_*(X) = T_{\eta\circ\sigma}(0) = (\frac{d(a_1 \circ \sigma)}{dt}(0),\ldots,\frac{d(a_n \circ \sigma)}{dt}(0))$$

$$= (Xa_1,\ldots,Xa_n) = \overline{D}_X N = L(X).$$

The map L is C^∞ on M in the sense that if X is C^∞ on the subset A of M then $L(X) = (Xa_1,\ldots,Xa_n)$ is also C^∞ on A since each a_1 is C^∞ on M.

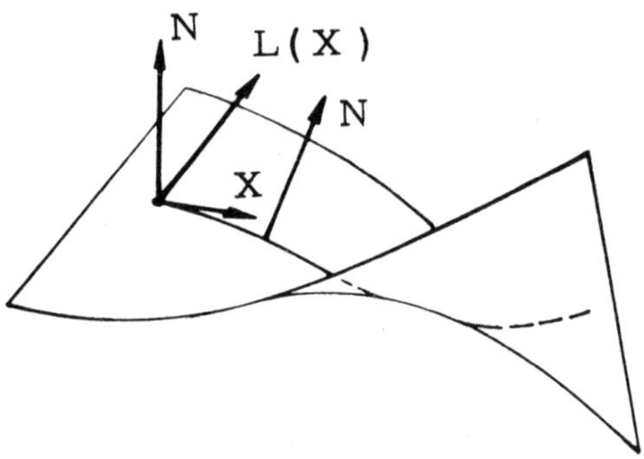

Fig. 2.1 The Weingarten Map (derivative of normal)

Our next objective is to show L is *self-adjoint* or *symmetric*; i.e., if X, Y are in M_p then $\langle L(X), Y\rangle = \langle X, L(Y)\rangle$.

To do this, let Z be a C^∞ field defined on a special coordinate neighborhood U of p and let $\overline{U}$ be the associated coordinate neighborhood of p in R^n with coordinate functions $\overline{x}_1,\ldots,\overline{x}_n$. Then $Z = \Sigma_{i=1}^{n-1} g_i(\partial/\partial x_i)$, where g_i are C^∞ real valued functions on U. We want to extend Z to a C^∞ field $\overline{Z}$ on $\overline{U}$, i.e., we want $\overline{Z}$ so that $\overline{Z}_p = Z_p$ for p in U. Let us assume the coordinate map ϕ maps $\overline{U}$ onto a ball, B, about the origin in R^n, i.e., $\overline{x}_i(p) = 0 = u_i \cdot\overline{\phi}(p)$ for all i. Then if $(t_1,\ldots,t_n)$ is in B, let π: $(t_1,\ldots,t_n) \to (t_1,\ldots,t_{n-1}, 0)$. This map π

(which is C^∞) induces a C^∞ map σ: $\bar{U} \to U$ by $\sigma = \bar{\phi}^{-1} \cdot \pi \cdot \bar{\phi}$. Letting $\bar{Z} = \Sigma_1^{n-1}(g_i \circ \sigma)(\partial/\partial\bar{x}_i)$, the field $\bar{Z}$ is a C^∞ extension of $\bar{Z}$ to $\bar{U}$.

Actually the above process allows us to extend an R^n-field Z that is C^∞ on U to a C^∞ field $\bar{Z}$ on $\bar{U}$.

Having the existence of such extensions we prove a proposition.

PROPOSITION. Let $\bar{U}$ and U be special neighborhoods of p as above and let $\bar{Z}$ and Z be C^∞ fields on $\bar{U}$ and U, respectively. Then $\bar{Z}$ is an extension of Z (i.e., $\bar{Z}_p = i_*(Z_p)$ for p in U) iff $(\bar{Z}f)|_U = Z(f|_U)$ for all f in $C^\infty(\bar{U}, R)$. If $\bar{X}$ and $\bar{Y}$ are C^∞ extensions to $\bar{U}$ of C^∞ fields X and Y on U, then $[\bar{X}, \bar{Y}]$ is a C^∞ extension of $[X, Y]$.

Proof. If $\bar{Z}_p = i_*(Z_p)$ for p in U, where i: $M \to R^n$ is the inclusion, then for f in $C^\infty(\bar{U}, R)$, $(\bar{Z}f)(p) = \bar{Z}_p f = (i_*(Z_p))f = Z_p(f \circ i) = Z(f|_U)(p)$. Conversely, if the two extreme terms are equal, then the second equality follows.

For the rest of the proposition, consider for p in U

$$[\bar{X}, \bar{Y}]_p f = \bar{X}_p(\bar{Y}f) - \bar{Y}_p(\bar{X}f) = X_p((\bar{Y}f)|_U) - Y_p((\bar{X}f)|_U)$$

$$= X_p(Y(f|_U)) - Y_p(X(f|_U)) = [X, Y]_p(f|_U),$$

thus $[\bar{X}, \bar{Y}]$ is an extension of $[X, Y]$.//

THEOREM. *The Weingarten map is self-adjoint.*

Proof. Take X and Y in M_p, imbed X and Y in C^∞ fields on a special neighborhood U of p, and extend X and Y to C^∞ fields $\bar{X}$ and $\bar{Y}$ on $\bar{U}$ as above. Then

$$\langle LX, Y\rangle - \langle X, LY\rangle = \langle \bar{D}_X N, Y\rangle - \langle X, \bar{D}_y N\rangle$$

$$= \langle \bar{D}_{\bar{x}}\bar{N}, \bar{Y}\rangle_p - \langle \bar{X}, \bar{D}_{\bar{y}}\bar{N}\rangle_p$$

$$= \bar{X}_p\langle \bar{N}, \bar{Y}\rangle - \langle \bar{N}, \bar{D}_{\bar{x}}\bar{Y}\rangle_p - \bar{Y}_p\langle \bar{N}, \bar{X}\rangle + \langle \bar{N}, \bar{D}_{\bar{y}}\bar{X}\rangle_p$$

$$= \langle \bar{D}_{\bar{y}}\bar{X} - \bar{D}_{\bar{x}}\bar{Y}, \bar{N}\rangle_p$$

$$= \langle [\bar{Y}, \bar{X}], \bar{N}\rangle_p = \langle [Y, X]_p, N_p\rangle = 0,$$

since $\bar{X}_p \langle \bar{N}, \bar{Y} \rangle = X_p \langle N, Y \rangle = 0 = Y_p \langle N, X \rangle$.//

The *fundamental forms* on M can now be defined in terms of L and the inner product. If X and Y are in M_p, then $I(X, Y) = \langle X, Y \rangle$, $II(X, Y) = \langle L(X), Y \rangle$, $III(X, Y) = \langle L^2(X), Y \rangle$, $IV(X, Y) = \langle L^3(X), Y \rangle$, etc., and these forms are called the *first, second, third, etc. fundamental forms* on M. Notice M is a Riemannian manifold with metric tensor defined by the first fundamental form. Since the inner product is symmetric and L is self-adjoint, the fundamental forms are all symmetric bilinear functions on $M_p \times M_p$ for all p in M. These forms are C^∞ in the sense that if X and Y are C^∞ fields with domain A, then $\langle L^k(X), Y \rangle_p = \langle L^k(X_p), Y_p \rangle$ is a C^∞ real valued function on A. The first three forms have a direct interpretation geometrically since L represents the Jacobian of the sphere map.

The algebraic invariants of the linear map L at each point now define the imbedded geometric invariants of the submanifold M at each point. Thus the determinant of L at p is the *total curvature* (*Gauss curvature*) $K(p)$ of M at p, the trace of L at p is the *mean curvature* $H(p)$, etc. The eigenvalues of L are the *principal curvatures* and the eigenvectors of L are the *directions of curvature* or *principal vectors*. Since L is self-adjoint there are always $(n - 1)$ independent directions of curvature. If L is a multiple of the identity map on M_p, then p is an *umbilic* point of M. If $L = 0$ at p we call p a *flat* point of M. Non-zero vectors X and Y in M_p are *conjugate* if $\langle LX, Y \rangle = 0$. A vector X(not zero) is *asymptotic* if it is self-conjugate, i.e., if $\langle LX, X \rangle = 0$. A curve in M is a *line of curvature* if its tangent is a principal vector at each of its points.

The following facts come immediately from these definitions. An asymptotic direction X is a direction of curvature iff $LX = 0$ iff X is conjugate to all vectors. Conjugate directions always exist since if $LX \neq 0$ then there exists a Y which is orthogonal to LX. If the second fundamental form $\langle LX, Y \rangle$ is positive or negative definite no asymptotic directions exist. If X and Y are two directions of curvature belonging to unequal eigenvalues, then X is orthogonal to Y. The proof of this is standard algebra, i.e.,

$$0 = \langle LX, Y \rangle - \langle X, LY \rangle = \langle k_1 X, Y \rangle - \langle X, k_2 Y \rangle = (k_1 - k_2)\langle X, Y \rangle,$$

so $k_1 \neq k_2$ implies $\langle X, Y\rangle = 0$. If X and Y are non-zero independent vectors with $LX = kX$ and $LY = -kY$, then the vectors $X + Y$ and $X - Y$ are orthogonal asymptotic directions spanning the same subspace as X and Y. Finally one notices that L must satisfy its characteristic polynomial, which will also give a relation between the fundamental forms, i.e., if $n = 3$, then $L^2 - HL + K(\text{identity}) = 0$ and $III - HII + KI = 0$.

When X is a principal vector, the Weingarten map says $\bar{D}_X N = kX$, where k is a principal curvature, and this equality is classically called the *formula of Rodrigues*.

Another classical concept is the *Dupin indicatrix* at each p in M which is the subset of M_p consisting of all vectors X such that $\langle L(X), X\rangle = \pm 1$.

Let $n = 3$ and let X and Y be unit orthogonal principal vectors in M_p with $LX = kX$ and $LY = hY$. If $Z = aX + bY$, then $\langle LZ, Z\rangle = ka^2 + hb^2$. Thus the indicatrix is the curve (or curves) in M_p such that $ka^2 + hb^2 = \pm 1$. Consider the three cases:

(1) If $K(p) > 0$, then h and k have the same sign (for $K = hk = \det L$) so suppose they are positive. The indicatrix is then an ellipse determined by $ka^2 + hb^2 = 1$, and p is an *elliptic point*.

(2) If $K(p) < 0$, then h and k have opposite signs, the indicatrix is two hyperbolas, and p is a *hyperbolic point*.

(3) If $K(p) = 0$, say $k = 0$, $h > 0$, then $b = \pm 1/\sqrt{h}$ gives two straight lines parallel to the X vector, and p is a *parbolic point*. (When $k = h = 0$, p is an *umbilic* and a *flat point*.)

There is a geometric interpretation of the indicatrix as an approximation to the intersection of the surface with a plane which is parallel and close to the tangent plane; for details see Struik (p. 84).

Section 2.3 The Gauss equation.

As in the last section, let M be a hypersurface of R^n, let $\bar{D}$ be the natural connexion on R^n, let N be a unit normal field that is C^∞ on M, and let $L(X) = \bar{D}_X N$ for X tangent to M. Let U and $\bar{U}$ be special coordinate neighborhoods of a point p in M and R^n respectively, and let $\bar{Z}$ be a C^∞ extension to $\bar{U}$ of a C^∞ field Z on U as usual.

If Y is a C^∞ field about p in M, and X in M_p, define $D_X Y$ by

$$(8) \qquad D_X Y = \bar{D}_X Y - \langle LX, Y\rangle N.$$

This is the *Gauss equation*. First notice $D_X Y$ is in M_p for

$$\langle D_X Y, N\rangle = \langle \bar{D}_X Y, N\rangle + \langle \bar{D}_X N, Y\rangle = X\langle Y, N\rangle = 0.$$

since $\langle Y, N\rangle = 0$ in a neighborhood of p. Next notice if X, Y are C^∞ on U, then $\bar{D}_X Y = \bar{D}_{\bar{X}} \bar{Y}|_U$ and $\langle LX, Y\rangle N$ are both C^∞ on U, so $D_X Y$ is C^∞ on U; because of this, we say D *is* C^∞.

Thus D becomes a candidate to define a *covariant differentiation* or a *connexion* on the submanifold M which is defined very simply from the natural connexion on R^n by decomposing $\bar{D}_X Y$ into its unique tangent and normal components relative to the tangent space of M. One must now check if the properties (1), (2), (3), and (4) are satisfied for D, and indeed they are, since they are satisfied for $\bar{D}$ and the second fundamental form is bilinear. The properties (5) and (6) are also valid for D, so D is the natural Riemannian connection associated with the induced metric (first fundamental form) on M (see Chapter 6). The proof of the first four properties is left to the reader, but we now show (5) and (6). Let Y and Z be fields on a neighborhood U about p, let $\bar{Y}$ and $\bar{Z}$ be extensions to $\bar{U}$, and let X be in M_p. Then

$$(D_Y Z - D_Z Y)_p = (\bar{D}_Y Z - \bar{D}_Z Y)_p = (\bar{D}_{\bar{Y}} \bar{Z} - \bar{D}_{\bar{Z}} \bar{Y})_p$$

$$= [\bar{Y}, \bar{Z}]_p = [Y, Z]_p,$$

and

$$X\langle Y, Z\rangle = X\langle \bar{Y}, \bar{Z}\rangle = \langle \bar{D}_X \bar{Y}, \bar{Z}\rangle + \langle \bar{Y}, \bar{D}_X \bar{Z}\rangle$$

$$= \langle D_X Y, Z_p\rangle + \langle Y_p, D_X Z\rangle.$$

Thus the natural metric tensor and connexion on R^n induce a Riemannian metric and Riemannian connexion on the hypersurface M.

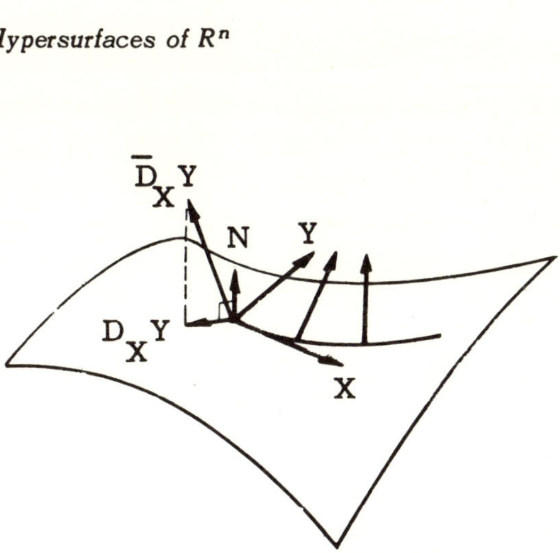

Fig. 2.2 **Decomposition of $\bar{D}_X Y$**

Since the Gauss equation induces a connexion D on M, one can define parallel vector fields along a curve and geodesics exactly as in section 2.1. If σ is a C^∞ curve in M with tangent T and Y is a C^∞ field along σ, then Y is *parallel along* σ if $D_T Y = 0$ along σ. The curve σ is a *geodesic* if $D_T T = 0$ along σ.

Application of the Gauss equation to the tangent field along a curve gives two results immediately.

THEOREM. *Let M be a hypersurface in R^n. A curve in M is a geodesic in R^n iff it is an asymptotic geodesic in M. A curve in M, which is not a geodesic in R^n, is a geodesic in M iff $\bar{D}_T T$ is normal to M along the curve (whose tangent is T).*

Proof. Let g be a curve in M with tangent T. The Gauss equation implies $\bar{D}_T T = D_T T - \langle LT, T\rangle N$. Thus $\bar{D}_T T = 0$ iff $D_T T = 0$ and $\langle LT, T\rangle = 0$. And $D_T T = 0$ iff $\bar{D}_T T$ is normal to M.//

Corollary. If M_1 and M_2 are two hypersurfaces of R^n and g is a geodesic on both hypersurfaces that is not a geodesic in R^n, on any parameter interval, then M_1 and M_2 are tangent along g (i.e., their tangent spaces coincide along g).

Proof. Let T be the tangent to g. Since $\bar{D}_T T \neq 0$ on any parameter interval, the normals to M_1 and M_2 determine the same subspace on a dense set of the parameter domain. Hence M_1 and M_2 are tangent along g.//

Section 2.4. The Gauss curvature and Codazzi-Mainardi equations.

Let M, N, L, D and $\bar{D}$ be as in the previous two sections. Our current goal is the "theorema egregium" of Gauss. This will show the "curvature" is independent of the imbedding, and motivate the definition of Riemannien curvature and curvature of a general connection.

Let X, Y, and Z be C^∞ fields on an open set A in M. Notice that

$$\bar{D}_X(\bar{D}_Y Z) - \bar{D}_Y(\bar{D}_X Z) - \bar{D}_{[X,Y]}Z$$

$$= (XYz_1,\dots,XYz_n) - (XYz_1,\dots,YXz_n) - ([X, Y]z_1,\dots,[X, Y]z_n) = 0$$

where $Z = (z_1,\dots,z_n)$ and z_i are C^∞ real valued functions on A. This fact will later verify that the "curvature of R^n is zero." By applying the Gauss equation and decomposing the above expression into tangent and normal parts, one obtains the Gauss curvature (9) and Codazzi-Mainardi (10) equations, respectively.

Thus,

$$0 = \bar{D}_X(D_Y Z - \langle LY, Z\rangle N) - \bar{D}_Y(D_X Z - \langle LX, Z\rangle N) - \bar{D}_{[X,Y]}Z$$

$$= D_X D_Y Z - \langle LX, D_Y Z\rangle N - X(\langle LY, Z\rangle)N - \langle LY, Z\rangle L(X)$$

$$- D_Y D_X Z + \langle LY, D_X Z\rangle N + Y(\langle LX, Z\rangle)N + \langle LX, Z\rangle L(Y)$$

$$- D_{[X,Y]}Z + \langle L([X, Y]), Z\rangle N.$$

Equating tangent and normal parts to zero gives

(9) $$D_X D_Y Z - D_Y D_X Z - D_{[X,Y]}Z = \langle LY, Z\rangle L(X) - \langle LX, Z\rangle L(Y)$$

and

$$\langle D_X L(Y) - D_Y L(X) - L([X, Y]), Z\rangle = 0$$

for all Z, so

(10) $$D_X L(Y) - D_Y L(X) - L([X, Y]) = 0.$$

Define $R(X, Y)Z = D_X D_Y Z - D_Y D_X Z - D_{[X,Y]}Z$, and notice (9) implies $R(X, Y)Z$ does not depend on the field nature of X, Y, and Z. Thus $R(X, Y)Z$ is a vector at p in A which depends only on X_p, Y_p and Z_p since these vectors are all that is needed to compute the left side of (9). Thus $R(X_p, Y_p)$ define a linear transformation on M_p called the *curvature of* X_p *and* Y_p. The justification of this definition is the following theorem, Gauss' "theorema egregium."

THEOREM. *Let* $n = 3$ *and let* X *and* Y *be an orthonormal base of* M_p. *Then the total curvature* $K(p) = \det L_p = \langle R(X, Y)Y, X\rangle$.

Proof. Using the Gauss curvature equation (9),

$$\langle R(X, Y)Y, X\rangle = \langle LY, Y\rangle\langle LX, X\rangle - \langle LX, Y\rangle\langle LY, X\rangle = \det L = K(p). //$$

The above theorem is significant because the term $\langle R(X, Y)Y, X\rangle$ depends only on the metric $\langle\ ,\ \rangle$ and the connexion D, and it is completely independent of the normal N or the map L. Thus the total curvature $K(p) = \langle R(X, Y)Y, X\rangle$ is an "intrinsic" invariant that is independent of the "imbedding" (i.e., of N and L). The theorem is generalized in Chapter 6.

Section 2.5. Examples.

See Figure 2.3 for sketches of (1), (2), and (3).

1. Let M be an $(n-1)$ dimensional hyperplane in R^n, i.e., let $N = (a_1,\ldots,a_n)$ determine a *constant* unit normal field on M. Then $L(X) = D_X N = (Xa_1,\ldots,Xa_n) = 0$ for all X at all points of M, i.e., $L \equiv 0$ on all of M. Thus M consists entirely of flat (umbilic) points, the total curvature K and mean curvature H (and all others) are identically zero. All the fundamental forms, except the first, are completely singular. Every victor is asymptotic and a direction of curvature, and all principal curvatures are zero.

2. Let M be S, the unit sphere about the origin in R^n, and let N be the outer normal on S, i.e., if $p = (a_1,\ldots,a_n)$ then $N(p) = (a_1, \ldots,a_n)$. Thus the sphere map η is the identity map, η_* is also the identity map, and hence $L(X) = X$ for all X. Thus $K \equiv 1$, $H \equiv (n-1)$ on S. All the fundamental forms are equal to the

first fundamental form, all points are umbilic, and all principal curvatures are unity. Every vector is a direction of curvature and there are no asymptotic directions.

3. Let M be the cylinder $C = [(t_1,\ldots,t_n) \text{ in } R^n\colon\ \Sigma_1^{n-1}(t_i)^2 = 1]$ with N = the "outer" normal. For $X = e_n = (0, 0,\ldots,0, 1)$ we have $LX = 0$, and for X orthogonal to e_n and tangent to C we have $LX = X$. Hence $K \equiv 0$, $H \equiv (n - 2)$, all principal curvatures are unity except one which is zero, etc.

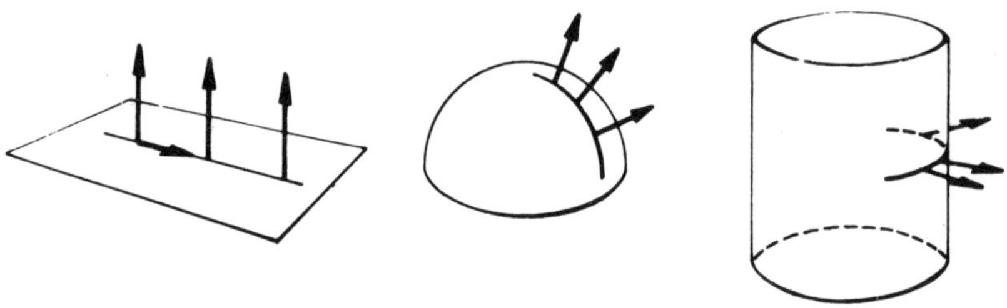

Fig. 2.3 Pieces of Examples (1), (2), (3)

4. Next let M be an open piece of a surface of revolution about the $z = e_3$ axis in R^3 (vaguely: M is obtained by revolving a C^∞ plane curve about an axis in the plane). Let P be a plane containing the z axis and take m in $M \cap P$ (and let us consider a point m not on the z axis at first).

Since the normal N lies in P, the vector $\bar{D}_X N = L(X)$ lies in P and is tangent to M so $L(X) = kX$ and X is a direction of curvature, where X is the unit tangent to a meridian curve. From the remarks preceding the examples there is a direction of curvature orthogonal to X, so the unit vector Y tangent to the parallel curves is a direction of curvature. The vector field $\bar{D}_X X$ is zero or orthogonal to X and must lie in the plane P, hence $\bar{D}_X X = \pm \bar{k}_1 N$, so $D_X X = 0$, and we see the meridians are geodesics. If the parallel curve through m is a geodesic, then $\bar{D}_Y Y$ is normal to M and not zero, since these curves are not geodesics in R^3. But $\bar{D}_Y Y$ is orthogonal to e_3, the z direction, hence a parallel curve is a geodesic on M iff the normal N along the parallel curve is horizontal (i.e., orthogonal to the z-axis). If m is a point on the z-axis, then every direction X is tangent to a meridian and hence is a direction of curvature, so m is umbilic and $K(m) \geq 0$.

5. Let us apply the analysis of example 4 to a torus, i.e., let M be obtained by rotating a circle C in the x, z-plane about the z-axis where we assume the circle does not intersect the z-axis. Then the meridians generated by C are geodesic, as is

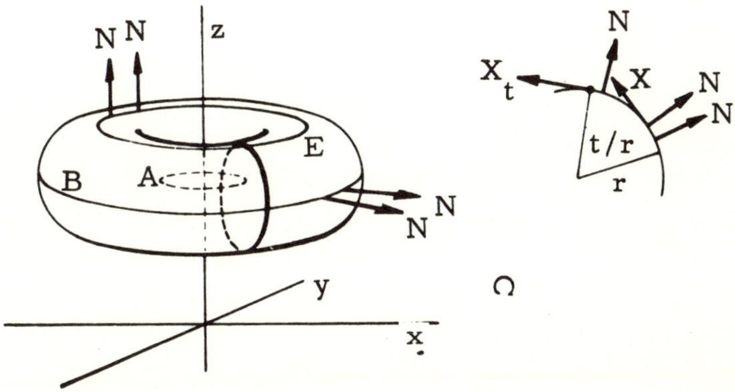

Fig. 2.4 The Torus

the minimum length parallel A and maximum length parallel B. Along B, M has positive curvature, along A the curvature is negative, and the curvature is zero on the extreme top and bottom curves E and F where N is constant. Indeed, if r_1 is the radius of A and r_2 the radius of B, then $a = (1/2)(r_2 - r_1)$ is the radius of C and

$$K = \frac{1}{ar_2} = \frac{2}{r_2(r_2 - r_1)} \quad \text{on } B,$$

$$K = -\frac{1}{ar_1} = \frac{-2}{r_1(r_2 - r_1)} \quad \text{on } A,$$

These expressions can be derived as follows. Let X be the unit tangent field to a circle of radius r about the origin in R^2 (see Figure 2.4) so $f(t) = (r \cos(t/r), r \sin(t/r))$ parameterizes the circle to fit X. Then evaluating a unit outer normal N on $f(t)$ gives $N \circ f(t) = (\cos(t/r), \sin(t/r))$. Hence, $(d/dt)(N \circ f(t)) = (1/r)X$, or if the circle lies on a surface then we see $\bar{D}_X N = L(X) = (1/r)X$. Now apply this to the circles on the torus.

6. We discuss ruled surfaces and developable surfaces briefly. A *ruled surface* is a two-dimensional submanifold M of R^3 such that through each point p in M there passes a segment of a straight line (the generator through p) which lies in M. When the normal field is a parallel field in R^3 along the generators, (thus the tangent plane is constant along generators), then the ruled surface is a *developable surface*. Notice we only consider the C^∞ case, although the above definitions can be generalized.

Let M be a ruled surface, and let X be a C^∞ unit vector tangent to the generator at each point of M. The generators are geodesics in R^3, so $\bar{D}_X X = 0$, and hence, from the Gauss equation, $D_X X = 0$ and $\langle LX, X\rangle = 0$ (so generators are asymptotic lines). Let Y be a unit vector field orthogonal to X in the neighborhood of a point p, then $K = \langle LX, X\rangle\langle LY, Y\rangle - \langle LX, Y\rangle^2 = -\langle LX, Y\rangle^2 \leq 0$ in this neighborhood. *Thus a ruled surface has non-positive curvature.* For a *defelopable surface,* $0 = \bar{D}_X N = LX$ so $K \equiv 0$. A theorem due to Massey (see Chapter 3) states a closed connected surface is developable iff $K \equiv 0$.

We study the neighborhood of a point p in a ruled surface M. Let $f(t)$ be the C^∞ curve through p which is parameterized by arc-length and is orthogonal to the generators at each point. Let T be the tangent to f (say $T = Y$ along f), and let $f(0) = p$. Then the map $(t, s) \to f(t) + sX(t)$ gives a coordinate system from a neighborhood of $(0, 0)$ in R^2 to a neighborhood of p in M.

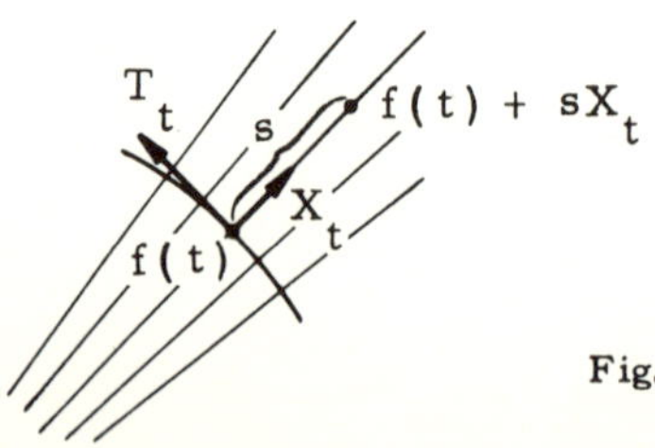

Fig. 2.5 Ruled Surface

Let N be a local unit normal for this coordinate neighborhood. The unit fields X, T, N give an orthonormal frame along f, and we next obtain the Frenet formulas for this frame. On f we have

$$1 = \langle X, X\rangle = \langle T, T\rangle = \langle N, N\rangle \text{ so } 0 = T\langle X, X\rangle = 2\langle \bar{D}_T X, \ X\rangle$$

implies $\bar{D}_T X$ normal to X. Similarly, $\bar{D}_T N$ normal to N and $\bar{D}_T T$ normal to T. Thus we define functions $a(t)$, $b(t)$, $c(t)$ by

$$\bar{D}_T T = aX + bN$$

$$\bar{D}_T X = -\,aT + cN$$

$$\bar{D}_T N = -\,bT - cX,$$

where $a = \langle \bar{D}_T T, X\rangle = T\langle T, X\rangle - \langle T, \bar{D}_T X\rangle = -\langle T, \bar{D}_T X\rangle$, etc. Holding s constant, we get a curve $f_s(t) = f(t) + sX(t)$ on M with tangent

$$A = T + s\bar{D}_T X = (1 - as)T + scN$$

(note this $T(t)$ and $N(t)$ are vectors at $f(t)$ which are rigidly translated in R^3 to $f_s(t)$ to give $A(t)$). The tangent space along a generator is spanned by A and X (and A is orthogonal to X), hence this tangent space is *constant* along a generator iff $c = 0$. The function $c/(c^2 + a^2)$ is called the *distribution parameter* and it is independent of the particular orthogonal trajectory f (which we show later). Thus (a) M is developable, (b) $K = 0$, (c) $c = 0$, (d) $LX = 0$, (note $\langle LX, T\rangle = \langle LT, X\rangle = -\,c$), and (e) $\bar{D}_T X$ is tangent to M, are all equivalent for M closed and connected (assuming Massey's theorem).

Assuming M is closed (and ruled with $c \neq 0$), on each generator there exists a distinguished point called the *central point*, and these points determine the *curve of striction* on the surface. Fixing two generators, say for $t_1 < t_2$, we compute the length $J(s)$ of an orthogonal trajectory between these two generators by

$$J(s) = \int_{t_1}^{t_2} (\langle A, A\rangle)^{1/2} dt$$

$$= \int_{t_1}^{t_2} (1 - 2as + a^2s^2 + c^2s^2)^{1/2} dt$$

Let us find the value of s which minimizes $J(s)$, and we get $J'(s) = 0$ if

$$-2a + 2(a^2 + c^2)s = 0$$

or $s = a/(a^2 + c^2)$ at t_1 as $t_2 \to t_1$. Hence the *curve of stricture* is the curve

$$f + (a/(a^2 + c^2)X$$

as a function of t. This is precisely the point on each generator where the tangent plane is normal to $\bar{D}_T X(t)$ since $\bar{D}_T X$ is orthogonal to X we know, and $0 = \langle \bar{D}_T X, A \rangle = -a + a^2s + c^2s$ again gives $s = a/(a^2 + c^2)$. As a problem we leave the formula for the curvature,

$$K(t, s) = -c^2/(1 - 2as + a^2s^2 + c^2s^2)^2,$$

and hence the central point on each generator is also characterized as the point where K is a maximum ($|K|$ a minimum). At the central point, $K = -(a^2 + c^2)^2/c^2$, which shows the distribution parameter $c/(a^2 + c^2)$ depends only on the generator.

If $s = 0$ gives the central point on a particular generator, i.e., we take our orthogonal curve f from this central point, then $\bar{D}_T X$ is normal to M at $s = 0$ and $a = 0$. Thus the distribution parameter $p = 1/c$ and $K(t, s) = -c^2/(1 + c^2s^2)^2 = -p^2/(p^2 + s^2)^2$. Along this generator $A = T + scN$ where T and N are vectors at the central point, hence the normal $N(s)$ along the generator is given by

$$N(s) = (-scT + N)/(1 + s^2c^2)^{1/2}$$

$$= (-sT + pN)/(p^2 + s^2)^{1/2}.$$

Thus if ϕ is the angle between the normal $N(s)$ and the normal N at the central point, we have $\tan \phi = s/p$, i.e., *the tangent of ϕ is directly proportional to the distance from the central point.* This is Chasles theorem (1839). This also shows the tangent plane turns evenly through 180° along a generator (turning through 90° on either side of the central point). For references, see Struik (p. 189) and Willmore (p. 107).

We point out we could have viewed the ruled surface discussed above as being generated by the curve $f(t)$ and the field $X(t)$ along the curve. To generate surfaces in this way X need not be orthogonal to T. Indeed, in case $\bar{D}_T T \neq 0$, then we generate a surface via $(t, s) \to f(t) + sT(t)$, for small $s > 0$ (or small $s < 0$); which we call the *tangential developable* of the curve f, which is the *edge of regression* of these two surfaces. It is a surface, since $A = T + s\bar{D}_T T$ is independent of $X = T$ (for $s \neq 0$), and the tangent space along a generator will be determined by T and $\bar{D}_T T$ for all s; hence the surface is developable. It is, of course, not a closed surface in general (see. Struik, p. 66).

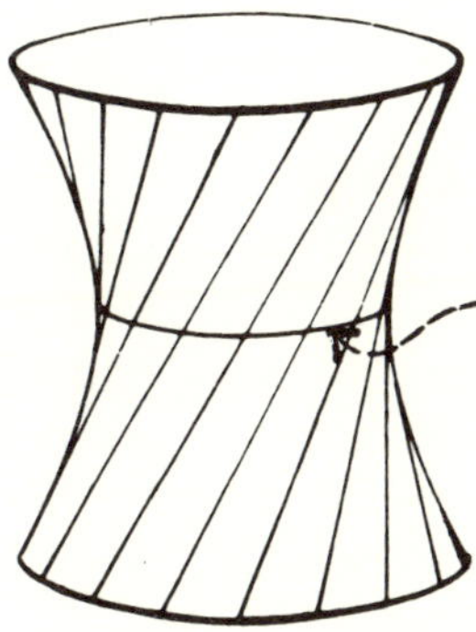

Fig. 2.6 Hyperboloid of Revolution

Section 2.6 Some applications.

Let M be a hypersurface of R^n with unit normal $N = (a_1,\ldots,a_n)$ where each a_i is a C^∞ function on M and $\Sigma_1^n a_i^2 = 1$. For any r in R, let $M_r = [p + rN_p\colon\ p \text{ in } M]$. Thus if $p = (p_1,\ldots,p_n)$ is in M, then $f(p) = p + rN_p = (p_1 + ra_1(p),\ldots,p_n + ra_n(p))$ is in M_r. The map f is called the *natural map* of M into M_r, and if f is univalent, then M_r is a *parallel hypersurface* of M with unit normal N, i.e., $N_{f(p)} = N_p$ for all p in M. Let L_r be the Weingarten map on N_r.

THEOREM. *Let* $f\colon M \to M_r$ *as just described. Then for* X *in* M_p, $f_*(X) = X + rL(X)$, $L_r(f_*X) = L(X)$, *and* f *preserves principal directions of curvature, umbilics, and the third fundamental form. Also*

$$\langle f_*X, f_*Y\rangle = \mathrm{I}(X, Y) + 2r\,\mathrm{II}(X, Y) + r^2\,\mathrm{III}(X, Y),$$

where I, II, III *are the first, second, and third fundamental forms on* M. *If* k *is a principal curvature of* M *at* m *in direction* X, *then* $k/(1 + rk)$ *is the corresponding principal curvature of* M_r *at* $f(m)$ *in direction* f_*X.

Proof. To compute f_*X, take a curve $\sigma(t) = (b_1(t),\ldots,b_n(t))$ with $X = (b_1'(0),\ldots,b_n'(0))$, and compute the tangent to $f\circ\sigma$ at $t = 0$. Let $N(\sigma(t)) = (a_1(t),\ldots,a_n(t)$; then $f\circ\sigma(t) = (\ldots,b_i(t) + ra_i(t),\ldots)$, and its tangent at $t = 0$ is indeed $X + rL(X)$. Also $N(\sigma(t)) = N(f\circ\sigma(t))$ from the definition of f and M_r. Thus $L(X) = \bar{D}_X N = (a_1'(0),\ldots,a_n'(0)) = \bar{D}_{f_*X}N = L_r(f_*X)$. This shows

$$\mathrm{III}_r(f_*X, f_*Y) = \langle L_rf_*X, L_rf_*Y\rangle = \langle LX, LY\rangle = \mathrm{III}(X, Y).$$

Now let X be a unit vector at m in M with $LX = kX$, so $L_r(f_*X) = LX = kX$ and $f_*X = (1 + rk)X$. If $1 + rk = 0$, then $f_*X = 0$ and $L_r(f_*X) = kX = 0$, so $k = 0$ and $1 = 0$, thus $1 + rk \neq 0$ if M_r is a hypersurface. Hence $L_r(f_*X) = (k/(1 + rk))f_*X$, which shows f preserves directions of curvature and umbilics. Finally, one can verify the expression for $\langle f_*X, f_*Y\rangle$ by direct computation using $f_*X = X + rLX$.//

Corollary. In the hypothesis of the above theorem let $n = 3$, and let the total curvature and mean curvature of M (and M_r) be denoted by K (and K_r) and H (and H_r). Then

$$K_r = K/(1 + rH + r^2K) \quad\text{and}\quad H_r = (H + 2rK)/(1 + rH + r^2K).$$

THEOREM. *Let* M *be a connected hypersurface in* R^n *consisting entirely of umbilics. Then* M *is either an open subset of a hyperplane or a sphere. If* M *is closed, then* M *is a hyperplane or a sphere.*

Proof. Take p in M and X_p in M_p, $X_p \neq 0$. Imbed X_p in a C^∞ field X about p and let Y be any other C^∞ field about p with X_p and

Y_p independent. Let $L = fI$ be the Weingarten map where f is a C^∞ real valued function on M and, I is the identity of each tangent space. By the Codazzi-Mainardi equation (10),

$$0 = D_X(fY) - D_Y(fX) - f[X, Y] = (X_p f)Y_p - (Y_p f)X_p,$$

since $D_X Y - D_Y X = [X, Y]$. The independence of X_p and Y_p implies $X_p f = 0$. Since M is connected, f must be a constant function on M (problem 14).

Suppose $L = kI$, k is constant on M. If $k = 0$, then $L \equiv 0$ on M, so N is constant on M ($\bar{D}_X N = 0$ for all X in M_p) and M must be an open subset of a hyperplane.

If $k \neq 0$, then we may assume $k > 0$ by changing the sign of N if necessary. Let $r = -1/k$ and let f: $M \to R^n$ by $f(p) = p + rN_p$. As in the preceding theorem, for all X in M_p, $f_*(X) = X + rL(X) = X - (1/k)kX = 0$. Thus $f_* = 0$, and since M is connected, f is a constant map. Let $c = p - (1/k)N_p$ for any p in M. Then all points of M are $1/k$ units from c. Thus M is an open subset of a sphere about c of radius $1/k$.//

Problems

23. Let f be in $C^\infty(R^2, R)$. Let M be the graph of f; thus $M = [(x, y, f(x, y))\colon (x, y) \text{ in } R^2]$. Let $W = [(f_x)^2 + (f_y)^2 + 1]^{1/2}$ and let $N = W^{-1}(-f_x, -f_y, 1)$. Show $X = (1, 0, f_x)$ and $Y = (0, 1, f_y)$ span M_m at all m and N is a unit normal that is C^∞ on M. Let $E = \langle X, X\rangle$, $F = \langle X, Y\rangle$, and $G = \langle Y, Y\rangle$. Show

$$L(X) = -W^{-3}[(f_{xx}G - f_{xy}F)X + (f_{xy}E - f_{xx}F)Y]$$

$$L(Y) = -W^{-3}[(f_{xy}G - f_{yy}F)X + (f_{yy}E - f_{xy}F)Y]$$

$$K = (f_{xx}f_{yy} - f_{xy}^2)/W^4$$

$$H = (-1/W^3)(f_{xx}G + f_{yy}E - 2f_{xy}F).$$

Compute $b_{11} = \langle LX, X\rangle$, $b_{12} = \langle LX, Y\rangle$, $b_{22} = \langle LY, Y\rangle$, $c_{11} = \langle LX, LX\rangle$, $c_{12} = \langle LX, LY\rangle$, and $c_{22} = \langle LY, LY\rangle$.

Show $K = (\det b_{ij})/(EG - F^2) = [(\det c_{ij})/(EG - F^2)]^{1/2} = (2F_{xy} - E_{yy} - G_{xx})/2W^4$. Show $L^2 - HL + KI = 0$ where I is the identity map. (Compare with section 3.4.)

24. Compute the invariants for the *right helicoid* (a ruled surface) which is the image of the map ϕ: $R^2 \to R^3$ defined by $\phi(u, v) = (u \cos v, u \sin v, av)$ for $a > 0$.

25. Show a curve on a surface is a line of curvature iff the surface normals along the curve form a developable surface (Monge.)

3. Surfaces in R^3

Throughout this chapter, M will denote a surface in R^3; i.e., M is a two-dimensional C^∞ submainfold of R^3. Let N be a C^∞ unit normal field on M (such an N always exists locally). Let $\bar{D}$ and D be the natural connexions on R^3 and M, respectively. Let $L(X) = \bar{D}_X N$ be the Weingarten map for X tangent to M. Let U be the set of umbilics on M and let $V = M - U$. Let K and H denote the Gauss curvature (total curvature) and mean curvature functions on M, respectively. Let h and k be the principal curvature functions on M where $h(p) \geq k(p)$ for all p in M. Thus $K(p) = \det L_p = h(p)k(p)$ and $H(p) = \text{trace } L_p = h(p) + k(p)$ for p in M.

Section 3.1. Smoothness and the neighborhood of a non-umbilic point.

The first theorm establishes the smoothness of the invariants of M and the local existence of C^∞ orthonormal principal vectors on V.

THEOREM 1. *The set of umbilics U is closed in M, so its complement V is open in M. The functions K and H are C^∞ on M. The functions $h = (H + \sqrt{H^2 - 4K})/2$ and $k = (H - \sqrt{H^2 - 4K})/2$ are C^0 on M and C^∞ on V. For any p in V there is a neighborhood A of p with $A \subset V$ and an orthonormal C^∞ base field of principal vectors on A.*

Proof. For any m in M, let B be the domain of a local coordinate system. By applying the Gram-Schmidt process to the coordinate vector fields on B, we obtain an orthonormal C^∞ base field Z, W on B. Since L is C^∞, the vectors $L(Z) = aZ + bW$ and $L(W) = bZ + cW$ are C^∞ on B, and hence the functions a, b, and c are C^∞ on B. Thus $K = ac - b^2$ and $H = a + c$ are C^∞ on B, and hence, on M.

The eigenvalues h and k must satisfy the algebraic equation $\lambda^2 - H\lambda + K = 0$ associated with the characteristic equation of L. Hence we get explicit global expressions for h and k by the quadratic formula, and they are clearly continuous, since they are the composite of continuous functions. The set U is precisely the set where $h = k$ or $H^2 - 4K = 0$, so by continuity, U is closed and V is open. Since $H^2 - 4K > 0$ on V, the functions h and k are C^∞ on V.

For p in V, let B, Z, and W be as in the first paragraph, with $B \subset V$. We distinguish two cases: (1) if $b(p) \neq 0$ and (2) if $b(p) = 0$. In case (1), choose the neighborhood $A \subset B$ such that $b \neq 0$ on A and let $Y' = bZ + (h - a)W$ and $X' = (a - h)Z + bW$. Then X', Y' are C^∞ orthogonal non-vanishing fields on A with $LY' = hY'$ and $LX' = kX'$. Let X and Y be unit fields in directions X' and Y', respectively. In case (2), suppose $a(p) > c(p)$, choose $A \subset B$ so $a > c$ on A, and let $Y' = (h - c)Z + bW$ and $X' = bZ + (c - h)W$, etc.//

In the next theorem we derive basic expressions for studying the neighborhood of a non-umbilic point.

THEOREM 2. *Let m be a non-umbilic point on M and let X and Y be an orthonormal C^∞ base field of principal vectors on the neighborhood A of m with $A \subset V$ and $LX = kX$, $LY = hY$ on A. Defining the C^∞ functions a and b on A by*

$$a = (Yk)/(h - k) \quad \text{and} \quad b = -(Xh)/(h - k),$$

then $D_XY = aX$, $D_YX = bY$, $D_XX = -zY$, $D_YY = -bX$,

$$[X, Y] = aX - bY, \text{ and}$$

$$K = kh = \frac{(X^2h - Y^2k)(h - k) - (Xh)(2Xh - Xk) + (Yk)(Yh - 2Yk)}{(h - k)^2}$$

on A.

Proof. Since $\langle X, X\rangle = 1$, $\langle Y, Y\rangle = 1$, and $\langle X, Y\rangle = 0$ on A, $0 = X\langle Y, Y\rangle = 2\langle D_XY, Y\rangle$ so $D_XY = aX$ for some C^∞ function a, which we compute below. Similarly, $D_XY = bY$ for some b. Also $0 = X\langle X, X\rangle = 2\langle D_XX, X\rangle$ and $0 = X\langle X, Y\rangle = \langle D_XX, Y\rangle + \langle X, D_XY\rangle$, so $D_XX = -aY$, and similarly, $D_YY = -bX$. Then $[X, Y] = D_XY - D_YX = aX - bY$.

To compute the expressions for a and b in terms of X, Y, h and k, we apply the Codazzi- Mainardi equation. Thus $D_X LY - D_Y LX = (Xh)Y + haX - (Yk)X - kbY = L([X, Y) = akX - bhY$. Equating coefficients of X and Y leads to the expressions for a and b.

To compute K, first notice $R(X, Y)Y = D_X(-bX) - D_Y(aX) - D_{(aX-bY)}Y = -(Xb)X - (Ya)X - a^2X - b^2X$. By the Gauss curvature equations, $K = \langle R(X, Y)Y, X\rangle = -(Xb) - (Ya) - a^2 - b^2$, and the final expression for K follows by inserting the formulas for a and b and computing.//

Corollary. If m is a non-umbilic critical point of both principal curvatures, then $K(m) = (X^2h - Y^2k)/(h - k)$. If M has no umbilics and K and H are constant (or the principal curvatures are constant), then $K = 0$.

Section 3.2. Surfaces of constant curvature.

Let M be a closed connected surface in R^3 with constant Gauss curvature K. Then M is a sphere, a developable surface, or doesn't exist, according as $K > 0$, $K = 0$, or $K < 0$, respectively. The cases when $K > 0$ (due to Liebmann) and $K < 0$ (due to Hilbert) were solved around 1900. It is amazing that the case $K = 0$ (due to Massey) was not completely solved until 1962.

Consider the case $K > 0$. The result of Liebmann follows from a lemma due to Hilbert.

LEMMA. *If K is a positive constant on M, then h cannot have a relative maximum (and k cannot have a relative minimum) at any non-umbilic point.*

Proof. Suppose m in V and m is a relative maximum for h and a relative minimum for k (since $K = hk =$ constant). With the notation of Theorem 2, $X^2h \leq 0$ and $Y^2k \geq 0$ at m. Thus by the above corollary, $K(m) \leq 0$, which is a contradiction.//

A theorem of Bonnet, proved in Chapter 10, shows the "compact" assumption in the following theorem can be replaced by "closed."

THEOREM 3. *A compact connected surface in R^3 of constant positive Gauss curvature is a sphere.*

Proof. At all points, the principal curvature $h \geq \sqrt{K}$, since $h^2 \geq hk = K$. Since M is compact, h must have an absolute maximum m in M, and m must be umbilic by Hilbert's lemma. Thus $h(m) = k(m) = \sqrt{K}$, and hence $h \leq \sqrt{K}$ on M. Thus $h = \sqrt{K}$, all points are umbilic, and M must be a sphere.//

The preceding theorem can be paraphrased by saying "a sphere cannot be bent." For a precise interpretation of this phrase, see Chapter 8, where a generalization, the rigidity theorem for convex bodies, is proved.

A proof of Hilbert's theorem stating that a closed connected surface with constant $K < 0$ cannot exist in R^3 is in Willmore. Here again, the compact case is easily disposed of by the first corollary of the following theorem; indeed, no compact M exists with variable $K \leq 0$ on M.

THEOREM 4. *On a compact surface in R^3 there is a point m with* $K(m) > 0$.

Proof. Let $r(p) = |p|$ give the distance from a point p in R^3 to the origin. Then $r \circ i$ is a continuous function on the compact surface M so it takes on a maximum at a point m in M. By a rotation (orthogonal transformation) of R^3, we may assume m lies on the z-axis (or u_3-axis). Let N be a C^∞ unit normal to M on a neighborhood of m with $N_m =$ $(0, 0, 1)$. Let X be any unit principal vector at m with $L(X) = \bar{D}_X N = kX$.

Let $\sigma(t) = (f(t), g(t), h(t))$ be a C^∞ curve on M with unit tangent vector X at $t = 0$; thus $X = (f'(0), g'(0), h'(0))$. Since m is an absolute maximum of $u_3 \circ i = z \circ i$ on M, $h''(0) < 0$. Letting X be the tangent to σ, we have at m, $\bar{D}_X X = (f''(0), g''(0), h''(0))$. Decomposing this vector into tangent and normal components, we get, by the Gauss equation, $-\langle LX, X\rangle N = (0, 0, -k) = (0, 0, h''(0))$, so $k = -h''(0) > 0$.

Since all principal curvatures are greater than zero at m, $K(m) > 0$.//

Notice the theorem is true for any compact hypersurface in R^n with a trivial modification of the proof.

Corollary. There is no compact hypersurface in R^n with non-positive Gauss curvature at all points.

Corollary. There is no compact minimal ($H = 0$) surface in R^3.

Proof. If $H = 0$, then $k = -h$ and $K = -h^2 \leq 0$.//

Before considering the case $K = 0$, recall that a generator on a surface M is a straight line in R^3 that lies on M with the normal to M constant along the line. A *developable* surface is a ruled surface with the normal constant along the ruling lines in the surface. If a developable surface is closed, then it has a generator through each point.

THEOREM 5. *Let M be a closed connected surface in R^3 with $K = 0$ on M. Then either M is a plane, or through each point of M passes a unique generator and all generators are parallel in R^3. Moreover, the mean curvature is constant along generators, and hence the boundary of the umbilic set is a union of these generators.*

Proof. Supposing M is not a plane; then the set V is non-empty. Let A be a connected neighborhood in V as described in theorems 1 and 2. Since H does not vanish on V and A is connected, we may assume $H = h > 0$ while $k = 0$ on A. Theorem 1 gives an orthonormal pair of C^∞ fields X and Y on A, with $LX = 0$ and $LY = HY$ on A. Since $Yk = 0$ on A, referring to theorem 2 we have $a = 0$ on A, so $D_X Y = 0$ and $D_X X = 0$ on A. By the Gauss equation, $\bar{D}_X X = D_X X - \langle LX, X\rangle N = 0$ on A. Thus the integral curves of X in A are straight line segments in R^3. Since M is closed, the continuation of these line segments must lie in M. Hence for p in V there is a unique line G_p through p with $G_p \subset M$. We next show $G_p \subset V$.

On the neighborhood A of p, by theorem 2,

$$K = 0 = \frac{X^2H}{H} - \frac{2(XH)^2}{H^2} = - HX^2(\frac{1}{H}).$$

Hence, if s is the arc length on G_p in the direction X with $s = 0$ at p, then $(1/H) = cs + d$ and $H = 1/(cs + d)$ for points in $G_p \,\hat{}\, A$. If there was an umbilic point at s' on G_p then $H(s') = 0$. At $s'' = \inf [s'\colon\ s' \text{ is umbilic}]$, $H(s'') = 1/(cs'' + d) \neq 0$, since H is continuous. Hence there are no umbilics on G_p, $G_p \subset V$, and to avoid an impossible singularity in H at $s = -c/d$, it follows H is constant on G_p.

After extending X and Y along G_p by letting X be the unit tangent to G_p, an overlapping neighborhood argument will show X and Y remain principal vectors; hence $L(X) = 0$ and $L(Y) = HY$ on G_p. Then $\bar{D}_X N = L(X) = 0$ implies N is constant on G_p, so G_p is a generator.

In the neighborhood A, since H is constant in the X direction, by theorem 2, $D_Y X = 0$, and so $\bar{D}_Y X = D_Y X - \langle LX, Y\rangle N = 0$. Thus X is parallel in R^3 along an integral curve of Y, which implies all generators through points in A are parallel. This implies all generators in one connected component of V must be parallel by another overlapping neighborhood argument. Hence the boundary of one connected component of V consists of two (or just one) lines parallel to the generators in that component. Consider now a connected component U_1 of the umbilic set. If U_1 has a non-empty interior in M, then this interior is an open surface of umbilics with $K = 0$, and hence it is an open subset of a plane in R^3. This open plane subset is bounded by two generator lines in the boundary of V, and these generator lines cannot intersect (by the uniqueness of the generators through points in V and its boundary), and hence they are parallel. Thus parallel generators are defined through all points of M.//

Corollary. A closed connected surface is a developable surface iff its Gauss curvature is identically zero.

Problem 26 provides additional theorems leading to surfaces with constant K and H, and it is hoped that by now their proofs would provide little difficulty. Another "classic" type of argument is provided by the following theorem and some of the theorems in the next section.

THEOREM 6. *Let M be a closed connected surface whose sphere map (Gauss map) is strictly conformal. Then M is a sphere or a minimal surface with negative curvature. If M is compact, it must be a sphere.*

Proof. Let η: $M \to S$ be the sphere map. Sincd η is strictly conformal, there is a C^∞ positive real valued scale function F on M with $\langle \eta_* X, \eta_* Y\rangle = \langle LX, LY\rangle = F(m)\langle X, Y\rangle$ for all X, Y in M_m for all m in M. Hence $\langle L^2(X) - F(m)X, Y\rangle = 0$ for all Y so $L^2(X) = FX$ for all X. One always has $L^2 - HL + KI = 0$, where I is the identity map; hence $HL = (K + F)I$. If $H(m) \neq 0$, then m is an umbilic and $K(m) = H^2(m)/4 > 0$. If m is umbilic and $H(m) = 0$, then $K(m) = -F(m) < 0$, but at an umbilic $K(m) = k^2(m) \geq 0$ always. Thus the umbilic set U is exactly the set of m where $H(m) \neq 0$, and hence U is open and closed. Since M is connected, either $M = U$ and M is a sphere ($F > 0$ rules out a plane) or $M = V$, $H = 0$, and $K = -F < 0$.

The last assertion of the theorem follows from a corollary to theorem 4.//

Section 3.3. *Parallel surfaces (normal maps).*

Let us state a standard hypothesis for some theorems (and problems) on "parallel surfaces": M is a closed connected surface in R^3 with C^∞ unit normal N, r is a non-zero real number, and f is the map f: $M \to R^3$ defined by $f(p) = p + rN_p$ (see section 2.6).

THEOREM 7. *With the standard hypothesis, if f is strictly conformal, then M is a sphere, plane, or has constant mean curvature $H = -2/r$ with no umbilics.*

Proof. From section 2.6, if X in M_m, then $f_*(X) = X + rL(X)$. Since f is strictly conformal, there is a C^∞ real valued function F on M with

$$<f_*X,\ f_*Y> = F(m)<X,\ Y> = <X + 2rLX + r^2L^2X,\ Y>$$

for all X, Y in M_m for all m in M. Hence, $r^2L^2 + 2rL + (1 - F)I = 0$ and, as always, $L^2 - HL + KI = 0$, so

$$(H + 2/r)L = [K - (1 - F)/r^2]I.$$

If $H(m) + 2/r \neq 0$, then m is an umbilic, and, indeed, $U = [m$ in M: $H(m) \neq -2/r]$. For if m umbilic and $H(m) = -2/r = 2k$, then $k = -1/r$, $K = 1/r^2$, $K - (1 - F)/r^2 = F/r^2 = 0$, and so $F(m) = 0$, which is impossible. Thus $M = U$ or $M = V$, and the only possibilities give the conclusion of the theorem.//

THEOREM 8. *With the standard hypothesis, if f preserves the second fundamental form, then M is a plane.*

Proof. From section 2.6, for all X and Y in M_m,

$$<LX,\ Y> = <L_r f_* X,\ f_*Y> = <LX,\ Y + rLY>;$$

thus $<LX,\ rLY> = <X,\ rL^2Y> = 0$ for all X and Y, and hence $L^2 = 0$. Thus the principal curvatures are zero, $L = 0$, and M is a plane.//

Similar results are given as problems. The following theorem is due to Bonnet, and the examples in the next section show the hypothesis is not vacuous.

THEOREM 9. *Let M be a surface of constant positive Gauss curvature K with no umbilics. Let $r_1 = 1/\sqrt{K}$ and $r_2 = -1/\sqrt{K}$ define parallel sets M_1 and M_2, respectively. Then M_1 and M_2 are immersions of M which have constant mean curvature $\sqrt{K}$ and $-\sqrt{K}$, respectively. If M' is a surface with constant mean curvature H (non zero) and non-zero Gauss curvature, letting $r = -1/H$ yields a parallel set that is an immersion of M' with constant positive Gauss curvature H^2.*

Proof. The proof is a corollary to the formulas for H_r and K_r in section 2.6. The special assumptions avoid trivial cases (sphere or cylinder) and singularities.

For the first part, f_* is non-singular, since for principal vectors $f_*X = (1 + rk)X$ and $1 + rk = 1 \pm k/\sqrt{K} \neq 0$, since there are no umbilics. Then $H_1 = (H + 2\sqrt{K})/(2 + H/\sqrt{K}) = \sqrt{K}$, and similarly, $H_2 = -\sqrt{K}$.

For the second part, f_* is non-singular, since $1 + rk = 1 - k/H = 0$ would imply $k = H$, so the other principal curvature is zero and $K = 0$ contrary to the hypothesis. Then $K_r = K/(1 - 1 + K/H^2) = H^2$.//

Section 3.4. Examples (surfaces of revolution).

Some general methods for computations with "parameterized" surfaces are introduced in this section. Let A be an open set in R^2 and let ϕ: $A \to R^3$ be defined by the three real valued slot functions f, g, and h, so $\phi(u, v) = (f(u, v), g(u, v), h(u, v))$ for (u, v) in A. Write $T_u = (f_u, g_u, h_u)$, where $f_u = \partial f/\partial u$, $T_{uv} = (f_{uv}, g_{uv}, h_{uv})$, etc. Notice $T_u = \phi_*(\partial/\partial u)$ is the tangent to the u-parameter curves on $\phi(A)$. Let us assume $(T_u \times T_v) \neq 0$, where "$\times$" is the cross-product of advanced calculus; thus ϕ is an immersion of A into R^3. Let $N = (T_u \times T_v)/W$ with $W = |T_u \times T_v| \neq 0$ on A.

To compute the Weingarten map L associated with N, notice $L(T_u) = \bar{D}_{T_u}(N) = N_u$. Thus,

$$\langle L(T_u), T_u\rangle = \langle N_u, T_u\rangle$$

$$= W^{-1}\langle T_{uu} \times T_v) + (T_u \times T_{uv}) - W_u N, T_u\rangle$$

$$= -\langle T_{uu}, N\rangle.$$

Similarly, $\langle L(T_i), T_j\rangle = -\langle T_{ij}, N\rangle$ with obvious values of i and j.

In case T_u and T_v are *orthogonal,*

$$L(T_u) = -\frac{\langle T_{uu}, N\rangle}{\langle T_u, T_u\rangle} T_u - \frac{\langle T_{uv}, N\rangle}{\langle T_v, T_v\rangle} T_v$$

and similarly for T_v; hence,

$$H = -\frac{\langle T_{uu}, N\rangle}{\langle T_u, T_u\rangle} - \frac{\langle T_{vv}, N\rangle}{\langle T_v, T_v\rangle}$$

and

$$K = \frac{\langle T_{uu}, N\rangle\langle T_{vv}, N\rangle - \langle T_{uv}, N\rangle^2}{\langle T_u, T_u\rangle\langle T_v, T_v\rangle}$$

A little more computation is necessary to determine the matrix for L in terms of T_u and T_v when they are not orthogonal.

Specializing further, let f be a positive (at least C^2) function, and for $u > 0$ let

$$\phi(u, v) = (u \cos v, u \sin v, f(u))$$

define a "surface of revolution." Applying the above analysis, one sees directly that T_u and T_v are principal vectors and $K = f'f''/u[1 + (f')^2]^2$ where $f'(u) = df/du$. To find surfaces of constant curvature one must solve the differential equation $f'f'' = uK[1 + (f')^2]^2$, a task that is left to reader via several problems. For more details and pictures, see Struik.

Section 3.5. Lines of curvature.

In this section we place some results involving lines of curvature, i.e., curves whose tangent vectors are principal directions of curvature.

Definition. *A* triply orthogonal system of surfaces *in a neighborhood U of R^3 is a family of surfaces such that through each point of U there passes exactly three members of the family whose normals are mutually perpendicular.*

THEOREM 10 (Dupin). *Intersecting surfaces from a triply orthogonal system intersect along a line of curvature.*

Proof. Let S_1, S_2 and S_3 be mutually orthogonal families of surfaces with unit normals N_i, respectively. Let $L_i X = \bar{D}_X N_i$, as usual. The field N_3 is a tangent to the intersection of S_1 and S_2, so one must show N_3 is a principal direction on S_1 and S_2 or $L_i(N_3) = a_i N_3$ for $i = 1, 2$. This is equivalent to showing $L_i(N_3)$ is orthogonal to N_1 and N_2 for $i = 1, 2$. To be specific, consider $L_1(N_3)$. Since $L_1(N_3)$ is tangent to S_1, $\langle L_1 N_3, N_1\rangle = 0$. While $\langle L_1 N_3, N_2\rangle = \langle \bar{D}_{N_3} N_1, N_2\rangle = -\langle N_1, \bar{D}_{N_3} N_2\rangle = -\langle N_1, L_2 N_3\rangle = -\langle L_2 N_1, N_3\rangle$, since L_2 is self-adjoint. Thus by symmetry, as one cyclicly permutes the indices,

$$\langle L_1 N_3, N_2\rangle = -\langle L_2 N_1, N_3\rangle = +\langle L_3 N_2, N_1\rangle = -\langle L_1 N_3, N_2\rangle.$$

Hence $\langle L_1 N_3, N_2\rangle = 0$.//

Examples of triply orthogonal systems are given by the coordinate surfaces in rectangular coordinates, cylindrical coordinates, and spherical coordinates. Another example is provided by a system of confocal quadrics, i.e., the surfaces $\Sigma_1^3 (x_i)^2/(a_i - \lambda) = 1$, with $a_1 < a_2 < a_3$ fixed, are othogonal for unequal values of λ (see Struik, p. 100). The classic work in this area is by Darboux.

THEOREM 14 (Liouville). *A conformal diffeomorphism of R^3 onto R^3 maps spheres into spheres.*

Proof. Let S be a sphere. For p in S, take an orthogonal family of curves on S and use the normal direction to S to generate an orthogonal family of surfaces. Adding in the "parallel" surfaces to S, one obtains a triply orthogonal system about p. Let f be the map in question, so f maps a neighborhood of p into a triply orthogonal system of surfaces about $f(p)$ on $F(S)$. By Dupin's theorem, the images of our original family of curves on S must be lines of curvature on $f(S)$. But we may choose an orthogonal family of curves on S to pass through any orthonormal pair of vectors X and Y at p. Hence all vectors tangent to $f(S)$ at $f(p)$ are principal, and $f(p)$ is an umbilic of $f(S)$. Thus $f(S)$ is completely umbilic, and since it is compact and connected it must be a sphere.//

The differentiability hypothesis in the above theorem is much too strong. The theorem can be used to show a conformal map of R^3 onto R^3 is a combination of similarities and isometries (also due to Liouville). For more details see Guggenheimer, p. 225.

We next discuss the behavior of the normal lines (in R^3) to a surface M along a line of curvature C. Let k be the principal curvature of M along C with respect to unit normal field N, and let X be a unit tangent to C. If $k \equiv 0$ on C, then $\bar{D}_X N = LX = kX = 0$ implies N is a constant field (in R^3) along C, and C is a plane curve (see section 6.3); thus the normal lines form a "cylinder," a developable surface. If k is a constant ($\neq 0$) along C, let $C(t)$ be the parameterization of C by arc length in the direction X, so $X(t) = C'(t) = (d/dt)C$. Then $kX = kC' = \bar{D}_X N = N'$, so $(N - kC)' = 0$ and $N(t) - kC(t) =$ constant vector. By translating the base (origin) of the vector $C(t)$, we have $N(t) = kC(t)$; thus all the normal lines along the curve pass through a single point (and thus form a "cone"). If $k \neq 0$ and $k' \neq 0$ along C, then let $B(t) = C(t) + f(t)N(t)$, so $B' = X + fkX + f'N$, and choosing $1 + fk = 0$ or $f(t) = -1/k(t)$, we obtain a curve B whose tangent developable gives the normal lines along C.

When both principal curvatures k and h are non-zero and non-constant in a neighborhood of p, then the points $p - (1/k)N$ and $p - (1/h)N$ are called the centers of principal curvature of p on M. The loci of the centers of principal curvature are called *center surfaces* (see Struik, p. 95).

Problems. All surfaces are in R^3.

26. If M is a closed connected surface with $K = 0$ and H constant, show M is a plane or a right circular cylinder. If M has no umbilics and K and H are constant, show M is a right circular cylinder. If $I = II$ or if $I = III$, show M is a sphere of radius one, and conversely. If $II = III$, show M is a sphere of radius one, a plane, or a right circular cylinder of radius one.

27. For $|u| < 1/b$ and $b > 0$, let $f(u) = \int_0^u bt/(1 - b^2t^2)^{1/2}dt$ and let ϕ: $(u, v) \to (u \cos v, u \sin v, f(u))$. Show that the surface of revolution determined by ϕ for $0 < u < 1/b$ is an open subset of of a sphere with curvature b^2.

28. For $0 < u < 1/b$ and $b > 0$, let $f(u) = \int_0^{b^{-1} \log bu} \sqrt{1 - e^{2bt}}dt$. Show that the surface of revolution induced by f has constant curvature $-b^2$ and draw its graph (tractrix).

29. Find a surface of constant positive curvature that is not an open subset of a sphere.

30. Show a surface is minimal ($H = 0$) iff there are orthogonal asymptotic vectors at each point.

31. Let $f(u) = \cosh^{-1}u$ for $u > 1$, and show the surface of revolution induced by f (catenoid) is a minimal surface ($H = 0$).

4. Tensors and Forms

The material in the first three chapters was based on a minimum amount of structure, i.e., manifolds, functions, and vector fields; moreover, there was a strong bias on hypersurfaces in Euclidean space. By this time the reader should be at home with these concepts, and before discussing general connexions on manifolds, it is convenient to define tensors and forms. They are there, and they are useful. At times in the past, one notices a strong compulsion to seek out and label tensors *ad nauseum*, and objects that were not tensors were eyed with suspicion. In a sense, this chapter is the "7th" section of Chapter 1; it is just more structure that a C^∞ manifold has automatically, and Chapter 7 continues the theme. It is hoped by breaking the definitions up they become more digestible.

Let M be a C^∞ n-manifold throughout this chapter, and let m be a point in M. Since the tangent space M_m at m is an n-dimensional vector space, the theory of linear algebra can be applied to define tensors and forms. A *p-covariant tensor at m* (for $p > 0$) or a *p-co tensor at m* is a real valued p-linear (i.e., linear in each slot) function on $M_m \times M_m \times \ldots \times M_m$ (p copies). Thus a is a 2-co tensor at m if

$$\alpha(X + Y, Z) = \alpha(X, Z) + \alpha(Y, Z)$$

$$\alpha(X, Y + Z) = \alpha(X, Y) + \alpha(X, Z)$$

$$\alpha(rX, Y) = \alpha(X, rY) = r\alpha(X, Y)$$

for all X, Y and Z in M_m and r in R. In a similar way, one defines a *V-valued p-co tensor at m*, where V is any vector space over R; indeed, V could be M_m itself.

Let M^*_m be the *dual space* of M_m. Thus M^*_m is the set of real valued 1-co tensors at m, or the set of linear functionals from M_m into R, and M^*_m is endowed with its natural vector space structure (i.e., one adds functions by adding their values and multiplies by a constant in an obvious way). Similarly, the set of p-co tensors at m, denoted by $T^{0,p}(M_m)$, is a vector space over R. A *p-contravariant* or *p-contra tensor at m* (for $p > 0$) is a real valued p-linear function on $(M^*_m)^p$, the cross product of p copies of M^*_m, and the natural vector space formed by p-contra tensors at m is denoted by $T^{p,0}(M_m)$. Define $T^{0,0}(M_m) = R$. (The sets of p-co tensor and p-contra tensors on any vector space W are denoted by $T^{0,p}(W)$ and $T^{p,0}(W)$, respectively.) Again, V-valued p-contra tensors are defined analogously. Finally, a *p-co and q-contra tensor at m* is a $(p + q)$-linear real valued function on $(M_m)^p \times (M^*_m)^q$, and the vector space of these tensors is denoted by $T^{q,p}(M_m)$. If p and q are greater than zero, elements of $T^{p,q}$ are called *mixed tensors*. Notice that a vector at m is a *1-contra tensor at m*. Similarly, there is a special name for a 1-co tensor at m, for it is called a *1-form at m*.

A tensor is *symmetric* iff its value remains the same for all possible permutations of its arguments (thus only $T^{p,0}$ or $T^{0,p}$ tensors can be symmetric). A tensor is *skew-symmetric* or *alternating* iff its value after any permutation of its arguments is the product of its value before the permutation and the sign of the permutation. For example, let α be a 3-co tensor at m and let π be a permutation of the set [1, 2, 3]. Then α is symmetric iff $\alpha^\pi(X_1, X_2, X_3) = \alpha(X_{\pi 1}, X_{\pi 2} X_{\pi 3}) = \alpha(X_1, X_2, X_3)$ for all permutations π and all vectors X_i in M_m. When α^π is defined by the first equality in the above line, α is alternating iff $\alpha^\pi = (-1)^\pi \alpha$, where $(-1)^\pi$ is the sign of the permutation π. Then a *p-form at m* (for $p > 0$) is an alternating p-co tensor at m, and the set of p-forms at m is denoted by $F^p(M_m)$. A 0-*form at m* is a real number; thus $F^0(W) = R$ for any vector space W over R. A p-form is said to be of *degree p*.

Tensor fields and C^∞ tensor fields are now defined in a way that is analogous to the definition of a vector field, once a vector was defined. For example, a *p-co tensor field* on a set U is a mapping that assigns to each m in U a p-co tensor at m. A *p-co tensor field α on a set U is* C^∞ iff U is open and for all sets of C^∞ vector fields $X_1,\dots,X_p$ on U, the function $[\alpha(X_1,\dots,X_p)](m) = \alpha_m(X_1(m),\dots,X_p(m))$ is a C^∞ function on U. A $C^\infty p$-form field on an open set U is called a *differential p-form* on U.

The *tensor product* of covariant tensors is defined as follows: if α in $T^{0,p}(W)$ and β in $T^{0,q}(W)$, then $\alpha \otimes \beta$ is the element in $T^{0,p+q}(W)$ defined by

$$(\alpha \quad \beta)(X_1,\dots,X_{p+q}) = \alpha(X_1,\dots,X_p)\beta(X_{p+1},\dots,X_{p+q})$$

for all X_i in W. Notice that $(\alpha_1 + \alpha_2) \otimes \beta = (\alpha_1 \otimes \beta) + (\alpha_2 \otimes \beta)$, $\alpha \otimes (\beta_1 + \beta_2) = (\alpha \otimes \beta_1) + (\alpha \otimes \beta_2)$, and $(r\alpha) \otimes \beta = \alpha \otimes (r\beta) = r(\alpha \otimes \beta)$ for r in R. However, $\alpha \otimes \beta \neq \beta \otimes \alpha$ in general, but $(\alpha \otimes \beta) \otimes \gamma = \alpha \otimes (\beta \otimes \gamma)$. Thus the tensor product is bilinear and associative but not symmetric. The tensor product of contravariant tensors or mixed tensors is defined analogously, but the details are omitted since these products are rarely used in this study.

If α and β are forms of degree p and q, respectively, then the *exterior, wedge,* or *Grassman product* $\alpha \wedge \beta$ is defined to be the $(p + q)$-form $\alpha \wedge \beta = (1/p!\ q!)\Sigma(-1)^\pi(\alpha \otimes \beta)^\pi$, where the sum is taken over all permutations π of the set $[1, 2,\dots,p + q]$. In problem 35 there is an expression for $\alpha \wedge \beta$ that avoids division. Notice that $\alpha \wedge \beta = (-1)^{pq}\beta \wedge \alpha$, $\alpha \wedge (\beta_1 + \beta_2) = \alpha \wedge \beta_1 + \alpha \wedge \beta_2$, where β_i are forms of the same degree, and $(\alpha \wedge \beta) \wedge \gamma = \alpha \wedge (\beta \wedge \gamma)$ which is proved by using problem 35.

To continue the definitions in terms of the abstract vector space W over R, the *tensor algebra* $T(W)$ over W and the *Grassman algebra (exterior algebra)* $F(W)$ over W are defined as the weak direct sums $T(W) = \Sigma_{p,q \geq 0} T^{p,q}(W)$ and $F(W) = \Sigma_{p \geq 0} F^p(W)$. By a *weak direct sum*, $\Sigma_I M_i$, of modules over an index set I, one means the set of formal finite linear combinations of elements $m_1 + m_2 + \dots + m_k$ where each m_i in M_i; or more precisely, $\Sigma_I M_i = \{f$ in $\Pi_I M_i$: $f(i) = 0$ for almost all (all but a finite number) of elements i in $I\}$, and then one writes $f = m_1 + m_2+\dots+m_k$ if $f(i) = m_i$ for $i = 1,\dots k$ and $f(j) = 0$ for $j \neq 1,\dots,k$, (see Chevalley[2] and Jacobson for more details). The tensor multiplication and the exterior product can be extended distributively to $T(W)$ and $F(W)$, respectively, thus making them algebras over R.

If U is an open set in the manifold M, let $T^{p,q}(U)$ be the set of C^∞ p-contra and q-co tensor fields on U, and let $T(U)$ and $F(U)$ be defined analogously. On the other hand, let $\mathfrak{F}_u$ be the ring of C^∞ real valued functions on U and let $\mathfrak{X}_u$ be the $\mathfrak{F}_u$-module of C^∞ vector fields on U. Then the above definitions can be extended to define the $\mathfrak{F}_u$-modules $T^{p,q}(\mathfrak{X}_u)$ and $F^p(\mathfrak{X}_u)$ for p, $q \geq 0$, where $T^{0,0}(\mathfrak{X}_u) = F^0(\mathfrak{X}_u) = \mathfrak{F}_u$. The next theorem and its corollary are designed to illuminate the relation between $T(U)$ and $T(\mathfrak{X}_u)$. To accomplish this, let us define an open set V in M to be *framed* if there exists a C^∞ *base field* on V, i.e., a set of n C^∞ vector fields $e_1,\dots,e_n$ on V that are independent at each point of V.

THEOREM (characterization of C^∞ tensors). *If U is a framed open set in M, then $T^{p,q}(U)$ is isomorphic $T^{p,q}(\mathfrak{X}_u)$ in a natural way.*

Proof. Let $e_1,\dots,e_n$ be a C^∞ base field on U, and let $w_1,\dots,w_n$ be the dual C^∞ 1-forms on U (see problem 32). It is sufficient to illustrate the proof for $T^{0,p}$ where $p > 0$, since the other cases are analogous. Consider a in $T^{0,p}(\mathfrak{X}_U)$, and let

$$\bar{a} = \Sigma_{1\leq i_j\leq n} a(e_{i_1}, e_{i_2},\dots,e_{i_p})[w_{i_1} \quad w_{i_2} \quad \otimes\dots\otimes w_{i_p}],$$

be an element in $T^{0,p}(\mathfrak{X}_U)$ defined by

$$[\bar{a}(X_1,\dots,X_p)](m)$$

$$= \Sigma_{1\leq i_j\leq n} [a(e_{i_1},\dots,e_{i_p})](m)[w_{i_1}(X_1(m)w_{i_2}(X_2(m))\dots w_{i_p}(X_p(m)].$$

where X_i are C^∞ fields on U. Then $a = \bar{a}$ as elements of $T^{0,p}(X_U)$, for if X_i are in $\mathfrak{X}_U$, then the function

$$a(X_1,\dots,X_p) = a(\Sigma_{i_1=1}^n w_{i_1}(X_1)e_{i_1}, \Sigma_{i_2} w_{i_2}(X_2)e_{i_2},\dots.\Sigma_{i_p} w_{i_p}(X_p)e_{i_p})$$

$$= \Sigma_{1 \leq i_j \leq n} w_{i_1}(X_1) w_{i_2}(X_2) \ldots w_{i_p}(X_p) a(e_{i_1}, \ldots, e_{i_p}),$$

since a is multilinear over $\mathfrak{F}_U$ and each $w_j(X_i)$ is a function in $\mathfrak{F}_U$.

But $\bar{a}$ is an element of $T^{0,p}(U)$, and notice $[\bar{a}(X_1, \ldots, X_p)](m)$ *depends only on the vectors* $X_1(m), \ldots, X_p(m)$ *and not on the fields* $X_1, \ldots, X_p$. Thus the map $a \to \bar{a}$ defines an isomorphism of $T^{0,p}(\mathfrak{X}_U)$ onto $T^{0,p}(U)$.//

One can "roughly" paraphrase the above theorem by saying that an $\mathfrak{F}_U$-multilinear function on vector fields on U is actually a smooth piecing together of R-multilinear functions on M_m for each m in U.

Corollary. Let U be open in M. Let a be a map that assigns to each framed open set $V \subset U$ an element a_V in $T^{q,p}(\mathfrak{X}_V)$ with $\bar{a}_V = \bar{a}_W$ in $T^{q,p}(V \cap W)$ for all open framed V and W contained in U. Then there is a unique tensor a in $T^{p,q}(U)$ such that $a|_V = \bar{a}_V$ for each framed open $V \subset U$. Moreover, if m in U and $X_1, \ldots, X_p$ are in M_m while $z_1, \ldots, z_q$ are in M^*_m, then

$$(*) \qquad a_m(X_1, \ldots, X_p, z_1, \ldots, z_q) = [a_V(\bar{X}_1, \ldots, \bar{X}_p, \bar{z}_1, \ldots, \bar{z}_q](m),$$

for any C^∞ fields $\bar{X}_i$ and C^∞ forms $\bar{z}_j$ on any framed neighborhood V of m with $\bar{X}_i(m) = X_i$ and $\bar{z}_j(m) = z_j$.

Proof. Use (*) to define a_m at any m in U. If W is any other framed open neighborhood of m, then $a_m = (\overline{a_V})_m = (\overline{a_W})_m$, and one need only know the values of fields and forms at m in order to evaluate both of the tensors on the right.//

If the reader will become familiar with tensors and computations involving their linearity via some of the problems, then the above theorem and corollary should become more natural.

To close this chapter we study the maps on tensors induced by a C^∞ map f: $M \to M'$, where M is a $C^\infty n$-manifold and M' is a $C^\infty n'$-manifold. Because the Jacobian f_* maps vectors on M into vectors on M', it induces a map f^* of covariant tensors (and forms) on M' into covariant tensors (and forms) on M. If g is in $T^{0,0}(U') = \mathfrak{F}_{U'}$, for open U' on M', then $f^*(g) = g \circ f$ is a C^∞ real valued function in $\mathfrak{F}_U$

where $U = f^{-1}(U')$. If a is a p-co tensor at $f(m)$ in M', then $(f^*a)_m$ is the p-co tensor at m defined on $X_1,\ldots,X_p$ in M_m by

$$(f^*a)_m(X_1,\ldots,X_p) = a_{f(m)}(f_*X_1,\ldots,f_*X_p).$$

If a is C^∞ on the open set U' in M', then f^*a is C^∞ on the open set $f^{-1}(U')$ in M. In the next paragraph we prove this for a 1-form a and leave the other cases to the problems.

Let a be a C^∞ 1-form on U', let X be any C^∞ vector field on U, and we show $(f^*a)(X)$ is a C^∞ function on U. Take m in U, let x_1, $\ldots,x_n$ be a coordinate system about m with domain $V \subset U$, and let y_1, $\ldots,y_{n'}$ be a coordinate system about $f(m)$ with domain $V' \subset U'$. Define C^∞ functions a_i on V and b_j on V' by $X = \Sigma_1^n a_i(\partial/\partial x_i)$ and $a = \Sigma_1^{n'} b_j(dy_j)$, where $dy_r(\partial/\partial y_s) = \delta_{rs} = 0$ or 1, according as $r \neq s$ or $r = s$, respectively (see problem 32). Then on V,

$$(f^*a)(X) = \Sigma a_i(b_j \circ f)\,\frac{\partial(y_j \circ f)}{\partial x_i}$$

for $i = 1,\ldots,n$ and $j = 1,\ldots,n'$, and since the right side is a C^∞ function on V, $(f^*a)(X)$ is C^∞ on V, and hence f^*a is C^∞ on U.

Finally, one checks that

$$f^*(a_1 + a_2) = f^*a_1 + f^*a_2,\ f^*(\gamma_1 \otimes \gamma_2) = (f^*\gamma_1) \otimes (f^*\gamma_2),$$

and $f^*(\beta_1 \wedge \beta_2) = (f^*\beta_1) \wedge (f^*\beta_2)$, where a_i are tensors of the same degree, γ_i are any covariant tensors, and β_i are alternating covariant tensors. Thus $f^*\colon\ F(M') \to F(M)$ is a degree preserving exterior-algebra map of the C^∞ forms on M' into the C^∞ forms on M.

There are certain natural tensors on every manifold called *universal* tensors. These are mixed tensors that let the arguments "work on each other." For example, let I be the 1, 1-tensor $I(w, X) = w(X)$ for X in M_m and w in M_m^*. Another is the 2, 2-tensor $E(w_1, w_2, X_1, X_2) = w_1(X_1)w_2(X_2)$, etc.

The 1, 1-tensors, $T^{1,1}(W)$, over a vector space W have a natural interpretation, for there is a natural isomorphism of $T^{1,1}(W)$ with the group, $\mathrm{Hom}_R\,(W, W)$, of linear transformation of W into itself. If B

is in $T^{1,1}(W)$, then let $\overline{B}$ be the linear map $\overline{B}(Z_i) = \Sigma_{j=1}^n B(w_j, Z_i)Z_j$, where $Z_1,\ldots,Z_n$ is a base of W with dual base $w_1,\ldots,w_n$ of W^* (see problem 36).

Problems. In these problems, W is an n-dim real vector space and M is a C^∞ n-namifold.

32. Let $e_1,\ldots,e_n$ be a base of W. For $i = 1,\ldots,n$, let $w_i(e_j) = \delta_{ij}$, where $\delta_{ij} = 0$ if $i \neq j$ and $\delta_{ii} = 1$. Show $w_1,\ldots,w_n$ is a base of W^*, and for θ in W^*, $\theta = \Sigma_{i=1}^n \theta(e_i)w_i$.

33. Let $e_1,\ldots,e_n$ be a base of W, and let $w_1,\ldots,w_n$ be the dual base of $W(w_i(e_j) = \delta_{ij})$. If a in $T^{0,2}(W)$, show $a = \Sigma_{i,j=1}^n a(e_i, e_j)w_i \otimes w_j$; thus a is determined by its values on a basis. If $f_1,\ldots,f_n$ another base, let $a(e_i, e_j) = a_{ij}$, $a(f_i, f_j) = b_{ij}$ and $f_j = \Sigma_{i=1}^n c_{ij}e_i$. Show $b_{ij} = \Sigma_{r,s=1}^n c_{ri}c_{sj}a_{rs}$.

34. Show $T^{1,0}(W)$ is isomorphic to W. Show $T^{p,q}(W)$ has dimension $(p+q)n$. Show $F^p(W)$ has dimension $\binom{n}{p} = (n!)/p!(n-p)!$.

35. Let a in $F^p(W)$ and β in $F^q(W)$. If $X_1,\ldots,X_{p+q}$ in W, show $a \wedge \beta(X_1,\ldots,X_{p+q}) = \Sigma(-1)^\pi a(X_{\pi_1},\ldots,X_{\pi_p})\beta(X_{\pi_{p+1}},\ldots,X_{\pi_{p+q}})$, where the sum is over all *shuffle* permutations π for p and q, i.e., if $1 \leq i < j \leq p$ or $p+1 \leq i < j \leq p+q$, then $\pi_i < \pi_j$.

36. Show $T^{1,1}(W)$ is isomorphic to $\mathrm{Hom}_R(W, W)$, the set of all R-linear maps of W into W, via the above map $B \to \overline{B}$, and show this map is independent of the base Z_i. Show the universal tensor I in $T^{1,1}(W)$ corresponds to the identity map on W.

37. If $e_1,\ldots,e_n$ is a C^∞ base field on U in M, and $w_1,\ldots,w_n$ is the set of dual 1-forms on U, show each w_i is C^∞ on U.

38. Let f be in $C^\infty(M, R)$. For p in M and X in M_p, let $(df)_pX = Xf$. Show $(df)_p$ is a 1-form at p. If $x_1,\ldots,x_n$ is a coordinate system with domain U, show $dx_1,\ldots,dx_n$ is the dual base to $\partial/\partial x_1,\ldots,\partial/\partial x_n$ and $df = \Sigma_{i=1}^n(\partial f/\partial x_i)dx_i$ on U. Show df is C^∞ on M. Show $d(f+g) = df + dg$ and $d(fg) = fdg + gdf$ for g in $C^\infty(M, R)$. If $y_1,\ldots,y_n$ is also a coordinate system on U and $w = \Sigma_1^n a_i dx_i =$

$\Sigma_1^n b_j dy_j$, show $b_j = \Sigma_{i=1}^n a_i(\partial x_i/\partial y_j)$. If $a = \Sigma a_{ij} dx_i \quad dx_j = \Sigma b_{ij} dy_i \quad dy_j$, find expressions for a_{ij} in terms of b_{ij}. If g in $C^\infty(M, M')$, show $g^* \circ d = d \circ g^*$ on $F^q(M')$.

39. If $w_1,\dots,w_n$ is a base of M_p^*, show there is a coordinate system $x_1,\dots,x_n$ about p with $(dx_i)_p = w_i$ for all i.

40. Let M^* be the set of all ordered pairs (m, w) for m in M and w in M_m^*. Let π: $M^* \to M$ by $\pi(m, w) = m$. Let $x_1,\dots,x_n$ be a coordinate system on M with domain U, and define functions $q_1,\dots,q_n$, $p_1,\dots,p_n$ on $\pi^{-1}(U)$ by $q_j = x_j \circ \pi$ and $w = \Sigma p_j(m, w) dx_j$. With these coordinate functions, M^* becomes a C^∞ $2n$-manifold called the *cotangent bundle of M*. The *fundamental 1-form* W *on* M *is defined by* $W_{(m,w)} = \pi^* w$. Show $W = \Sigma p_i dq_i$ on $\pi^{-1}(U)$. Show dW is a non-degenerate 2-form; i.e., $dW(X, Y) = 0$ for all Y implies $X = 0$. The forms W and dW are fundamental in classical mechanics (see G. W. Mackey).

41. If X is tangent to $B(M)$ at $b = (m\colon\ e_1,\dots,e_n)$. let $\bar{w}_i(X)$ be real numbers such that $X = \Sigma \bar{w}_i(X) e_i$. Show w_i are C^∞ 1-forms on $B(M)$ (see problem 22).

5. Connexions

This chapter is a study of a general connexion on a C^∞ manifold, the concepts belonging to the connexion, and the different ways of defining the connexion. These connexions are historically called affine or linear connexions on a manifold. The generalization to connexions in principal fiber bundles is sketched in section 5.5, but these generalizations will not be focused upon in these notes.

Section 5.1. Invariant viewpoint.

The approach to connexions that follows is due to Koszul and is found in Nomizu[2] and the first chapter of Helgason. The definition was motivated in section 2.1.

Let M be a C^∞ n-manifold. A *connexion, infinitesimal connexion,* or *covariant differentiation* on M is an operator D that assigns to each pair of C^∞ fields X and Y, with domain A, a C^∞ field $D_X Y$, with domain A; and if Z is a C^∞ field on A while f is a C^∞ real valued function on A, then D satisfies the following four properties:

(1) $D_X(Y + Z) = D_X Y + D_X Z$

(2) $D_{(X+Y)}(Z) = D_X Z + D_Y Z$

(3) $D_{(fX)} Y = f D_X Y$

(4) $D_X(fY) = (Xf)Y + f D_X Y.$

These properties imply the vector $(D_X Y)_m$, at a point m in M, depends only on X_m and the values of Y on some curve that fits X_m. For, let $e_1, \ldots, e_n$ be a C^∞ base field about m, let $X_m = \Sigma_1^n a_i(m)(e_i)_m$ and $Y = \Sigma_1^n b_j e_j$ on the domain of the base field (intersected with domain of Y). Then

$$(D_X Y)_m = [D_X(\Sigma b_j e_j)]_m$$

$$= \Sigma_j [(X_m b_j)(e_j)_m + b_j(m) \Sigma_i a_i(m)(D_{e_i} e_j)_m].$$

Thus $a_i(m)$, $b_j(m)$, and $X_m b_j$ determine $D_X Y$ completely if the fields $D_{e_i} e_j$ are known (see section 5.2).

The existence of many manifolds with connexions has been illustrated by the natural induced connexions on hypersurfaces of R^n.

Let σ be a curve in M with tangent field T. A C^∞ vector field Y on σ is *parallel along* σ iff $D_T Y = 0$ on σ. The curve σ is a *geodesic* iff $D_T T = 0$ on σ. Thus a curve is a geodesic iff its tangent field is a parallel field along the curve. The following two theorems give the existence of parallel fields and geodesics. The domain of an index of summation is always $1, \ldots, n$ unless otherwise specified.

THEOREM. *Let σ be a curve on $[a, b]$ with tangent T. For each vector Y in $M_{\sigma(a)}$ there is a unique C^∞ field $Y(t)$ on σ such that $Y(a) = Y$ and the field $Y(t)$ is parallel along σ. The mapping $P_{a,t}$: $M_{\sigma(a)} \to M_{\sigma(t)}$ by $P_{a,t}(Y) = Y(t)$ is a linear isomorphism which is called parallel translation along σ from $\sigma(a)$ to $\sigma(t)$*

Proof. Let $x_1,\ldots,x_n$ be a coordinate system about $\sigma(a)$ with domain U, and let $X_1,\ldots,X_n$ be the associated coordinate fields. We define C^∞ functions Γ^i_{jk} on U by $D_{X_k}X_j = \Sigma_i\Gamma^i_{jk}X_i$. Let σ map the domain $[a, b_1]$ into U. If $Y(t)$ is a field on σ with domain $[a, b_1]$, then define functions $a_1(t)$ on this domain by $Y(t) = \Sigma a_i(t)X_i(\sigma(t))$. Let $g_i(t) = x_i \circ \sigma(t)$ on $[a, b_1]$, so $T(t) = \Sigma g_j'(t)X_j(\sigma(t))$, where $g_j'(t) = (dg_j/dt)$. If $Y(t)$ is parallel along σ, then

$$0 = D_TY = \Sigma_i[a_i'X_i + a_i\Sigma_j g_j'\Gamma^k_{ij}X_k].$$

Thus $Y(t)$ parallel along σ iff

$$(5) \qquad \frac{da_k}{dt} + \Sigma_{i,j}a_i\frac{dg_j}{dt}\Gamma^k_{ij} = 0$$

for $k = 1,\ldots,n$ and for t in $[a, b_1]$. The condition $Y(a) = Y$ defines n initial values $a_i(a)$, and the theory of ordinary differential equations then gives a unique set of C^∞ functions $a_i(t)$, satisfying the above equations on the whole domain $[a, b_1]$, since the equations are linear. This defines the parallel field $Y(t)$.

For t in $[a, b_1]$, the map $P_{a,t}$ is linear because of the linearity of the equations (5).

If t is any number in $[a, b]$, we obtain $P_{a,t}$ by covering the compact set $\sigma([a, t])$ by a finite number of coordinate neighborhoods and parallel translating through each neighborhood via solutions of the systems (5).//

THEOREM. *Let m be in M, X in M_m. Then for any real number b there exists a real number $r > 0$ and a unique curve σ, defined on $[b - r, b + r]$ such that $\sigma(b) = m$, $T_\sigma(b) = X$, and σ a geodesic.*

Proof. Using the notation of the above proof, we must find C^∞ functions $g_i(t)$ that satisfy the second-order differential system,

$$(6) \qquad \frac{d^2g_k}{dt^2} + \Sigma_{i,j}\Gamma^k_{ij}\frac{dg_i}{dt}\frac{dg_j}{dt} = 0$$

with initial conditions $g_i(b) = x_i(m)$ and $X = \Sigma g_i'(b)X_i$. The theory of ordinary differential equations provides us with the $r > 0$ and the functions $g_i(t)$.//

The existence and uniqueness theory of ordinary differential equations will actually give us much more than the conclusion of the above theorem. In particular, if we let $\sigma(t; m, X, b)$ be the curve provided by the theorem, then the mapping σ is actually C^∞ with respect to all its parameters t, m, X and b.

The *torsion tensor* of a connexion D is a vector valued tensor Tor that assigns to each pair of C^∞ vectors X and Y, with domain A, a C^∞ vector field $\mathrm{Tor}(X, Y)$, with domain A, by

$$\text{(7)} \qquad \mathrm{Tor}(X, Y) = D_X Y - D_Y X - [X, Y].$$

One checks easily that $\mathrm{Tor}(X, Y) = -\mathrm{Tor}(Y, X)$, $\mathrm{Tor}(X+Y, Z) = \mathrm{Tor}(X, Z) + \mathrm{Tor}(Y, Z)$, and $\mathrm{Tor}(fX,Y) = f\,\mathrm{Tor}(X, Y)$, where f in $\mathcal{F}_A$ and Z in $\mathfrak{X}_A$. Thus the value of $\mathrm{Tor}(X, Y)$ at a point m depends only on X_m and Y_m, and not on the fields X and Y, by the theorem at the end of Chapter 4. If more than one connexion enters the discussion, we write Tor_D for the torsion of the connexion D. If $\mathrm{Tor}_D \equiv 0$, then we say D is *symmetric*, or *torsion free*.

As far as we know, there is no nice motivation for the word "torsion" to describe the above tensor. In particular, it has nothing to do with the "torsion of a space curve".

The following definition of curvature has been motivated in section 2.4.

The *curvature tensor* of a connexion D is a linear transformation valued tensor R that assigns to each pair of vectors X and Y at m a linear transformation $R(X, Y)$ of M_m into itself. We define $R(X, Y)Z$ by imbedding X, Y, and Z in C^∞ fields about m and setting

$$\text{(8)} \qquad R(X, Y)Z = (D_X D_Y Z - D_Y D_X Z - D_{[X,Y]}Z)_m.$$

Again we check linearity over the ring of C^∞ functions as coefficients on the right to determine the tensor character of R. Here, $R(X, Y)Z = -R(Y, X)Z$, and if f is C^∞, then

$$R(fX, Y)Z = fD_XD_YZ - (Yf)D_XZ - fD_YD_XZ + (Yf)D_XZ - fD_{[X,Y]}Z$$

$$= fR(X, Y)Z.$$

Also

$$R(X, Y)(fZ) = D_X[(Yf)Z + fD_YZ] - D_Y[(Xf)Z + fD_XZ] - ([X, Y]f)Z$$

$$- fD_{[X,Y]}Z = (XYf)Z + (Yf)D_XZ + (Xf)D_YZ + fD_XD_YZ - (YXf)Z$$

$$- (Xf)D_YZ - (Yf)D_XZ - fD_YD_XZ$$

$$- ([X, Y]f)Z - fD_{[X,Y]}Z = fR(X, Y)Z.$$

The linearity of $R(X, Y)Z$ with respect to addition (in each of its variables) is trivial to check.

The tensor nature of the torsion and curvature will again be verified in section 5.3 with exhibition of the classical coordinate representations of these tensors.

The concept of a "connexion-preserving" map follows naturally. Let M and M' be C^∞ manifolds with connexions D and D', respectively. A C^∞ map f: $M \to M'$ is *connexion preserving* if $f_*(D_XY) = D'_{f_*X}(f_*Y)$ for all vectors X and fields Y. Note the right side is well-defined since f_*Y is a well-defined field on some curve that fits f_*X. A C^∞ map f: $M \to M'$ is *geodesic preserving* if $f \circ g$ is a geodesic in M' for each geodesic g in M. Trivially, a connexion-preserving map is geodesic preserving.

THEOREM. *Let f be a diffeomorphism of M onto M', and let D' be a connexion on M'. Then there is a unique connexion D on M for which f is connexion preserving.*

Proof. Take X in M_m and let Y be a field about m. Since f is a diffeo, f_*Y is a field about $f(m)$. Define $D_XY = f_*^{-1}(D'_{f_*X}f_*Y)$. The verification that D is a connexion is left as an exercise.

If every geodesic $g(t)$ can be extended so it is a geodesic for all t in R, then the connexion D is *complete*.

Section 5.2. Cartan viewpoint.

For local problems concerning a connexion, one can transform the properties of D to certain properties of differential forms. By using fiber bundles associated with a manifold, one can also study global problems via differential forms. We develop the local viewpoint here.

Let D be a connexion on an n-manifold M, and fix D and M throughout this section. Let U be a fixed open set (perhaps a coordinate domain) in M, and let $e_1,\dots,e_n$ be a fixed base field of independent C^∞ vectors on U. Let $w_1,\dots,w_n$ be the C^∞ 1-forms on U which are the dual base to $e_1,\dots,e_n$ at each point of U. Define n^2 *connexion 1-forms* w_{ij} on U which are associated with D and the base field by

$$(9) \qquad D_X e_j = \Sigma_{i=1}^n w_{ij}(X) e_i.$$

The w_{ij} are linear by property (2) of the connexion D, and w_{ij} are C^∞, since if X a C^∞ field on U, then $D_X e$ is a C^∞ field, so $w_{ij}(X) = w_i(D_X e_j)$ is a C^∞ function.

The torsion and curvature tensors may also be expressed via differential forms associated with the base field. Define 2-forms T_i and R_{ij} on U by

$$(10) \qquad T(X, Y) = \Sigma_{i=1}^n T_i(X, Y) e_i$$

$$(11) \qquad R(X, Y) e_j = \Sigma_{i=1}^n R_{ij}(X, Y) e_i$$

where the properties of an alternating tensor are checked for T_i and R_{ij} via the properties of T and R.

The forms w_i, w_{ij}, T_i, and R_{ij} are related by the *Cartan structural equations* which are equivalent to the definition of the torsion and curvature tensors. We merely express everything in terms of the base field. Let X and Y be C^∞ fields on U. Then,

$$T_i(X, Y) e_i = D_X Y - D_Y X - [X, Y]$$

$$= D_X(\Sigma w_j(Y) e_j) - D_Y(\Sigma w_j(X) e_j) - \Sigma w_j([X, Y]) e_j$$

$$= \Sigma(Xw_j(Y) - Yw_j(X) - w_j[X, Y])e_j$$

$$+ \Sigma(w_j(Y)w_{ij}(X) - w_j(X)w_{ij}(Y))e_i.$$

Equating components,

$$T_i(X, Y) - (\Sigma w_{ij} \wedge w_j)(X, Y) = Xw_i(Y) - Yw_i(X) - w_i[X, Y].$$

Since the expression on the left is a 2-form, so is the expression on the right (taken as a whole), and indeed, it is the *exterior derivative* dw_i *of* w_i evaluated on X and Y. With this motivation we define the *exterior derivative operator* d on 1-forms and functions (0-forms) as follows.

For a C^∞ function f with domain A, let $df(X) = Xf$; thus df is a C^∞ 1-form on A. Let w be any C^∞ 1-form with domain A. Then dw is a C^∞ 2-form with domain A, defined on C^∞ fields X, Y on A by

(12) $$dw(X, Y) = Xw(Y) - Yw(X) - w[X, Y].$$

We leave it to the reader to check that the right side is linear in each slot over the ring of C^∞ functions on A, and hence that $dw(X_m, Y_m)$ is defined for m in A independent of the fields X and Y.

If f is a C^∞ function on A, then $d^2f = d(df) = 0$. To see this, let X and Y be C^∞ fields on A; then,

$$d^2f(X, Y) = Xdf(Y) - Ydf(X) - df[X, Y]$$

$$= XYf - YXf - [X, Y]f = 0.$$

Also note that if $x_1, \ldots, x_n$ a coordinate system on A, then $dx_1, \ldots, dx_n$ is the dual base to $\partial/\partial x_1, \ldots, \partial/\partial x_n$, since $dx_i(\partial/\partial x_j) = \partial x_i/\partial x_j = \delta_{ij}$ (the Kronecker delta).

Now we can write the *first Cartan structural equation,*

(13) $$dw_i = -\Sigma_{i=1}^n w_{ij} \wedge w_j + T_i.$$

By a computation involving the definition of $R(X, Y)$, which is completely analogous to the above computation, one obtains the *second*

Cartan structural equation,

(14) $$dw_{ij} = -\Sigma_{k=1}^{n} w_{ik} \wedge w_{kj} + R_{ij}.$$

These equations provide an alternate proof of the tensor character of T and R, since they show that T_i and R_{ij} are 2-forms.

Section 5.3. Coordinate viewpoint.

Let U be a coordinate neighborhood, and let $X_1,\ldots,X_n$ be the coordinate base field associated with the system $x_1,\ldots,x_n$ on U. Then $w_i = dx_i$ and the associated forms w_{ij}, T_i, and R_{ij} define functions Γ^i_{jk}, T_{ijk}, and R^i_{jkh}, respectively, by

$$\Gamma^i_{jk} = w_{ij}(X_k) \qquad \text{so} \qquad w_{ij} = \Sigma_k \Gamma^i_{jk} dx_k,$$

$$T_{ijk} = T_i(X_j, X_k) \qquad \text{so} \qquad T_i = \Sigma_{jk} T_{ijk} dx_j \otimes dx_k,$$

$$R^i_{jkh} = R_{ij}(X_k, X_h) \qquad \text{so} \qquad R_{ij} = \Sigma_{kh} R^i_{jkh} dx_k \otimes dx_h.$$

From the structural equations, we have

$$T_{ijk} = (d^2x_i + \Sigma\Gamma^i_{rs} dx_s \wedge dx_r)(X_j, X_k) = \Gamma^i_{kj} - \Gamma^i_{jk},$$

since $d^2x_i = 0$, and

$$R^i_{jkh} = (dw_{ij} + \Sigma(\Gamma^i_{rs} dx_s) \wedge (\Gamma^r_{jt} dx_t))(X_k, X_h)$$

$$= X_k w_{ij}(X_h) - X_h w_{ij}(X_k) + \Sigma(\Gamma^i_{rk}\Gamma^r_{jh} - \Gamma^i_{rh}\Gamma^r_{jk})$$

$$= \frac{\partial \Gamma^i_{jh}}{\partial x_k} - \frac{\partial \Gamma^i_{jk}}{\partial x_h} + \Sigma(\Gamma^i_{rk}\Gamma^r_{jh} - \Gamma^i_{rh}\Gamma^r_{jk}),$$

which are the classical coordinate components of these tensors.

Section 5.4. *Difference tensor of two connexions.*

The reference for this section is Ambrose, Singer and Palais. Let M be a C^∞ manifold, and let D and $\bar{D}$ be *connexions* on M. For fields X and Y we define the *difference tensor* $B(X, Y) = \bar{D}_X Y - D_X Y$. The linearity of B in the first slot is trivial from properties of the connexions (namely, (2) and (3)). To check the second slot, let f be C^∞ on the domain of X and Y; then $B(X, fY) = (Xf)Y + f\bar{D}_X Y - (Xf)Y - fD_X Y = fB(X, Y)$.

Let $B(X, Y) = S(X, Y) + A(X, Y)$ be the standard decomposition of a bilinear tensor into symmetric and skew-symmetric pieces; i.e.,

$$S(X, Y) = (1/2)[B(X, Y) + B(Y, X)]$$

and

$$A(X, Y) = (1/2)[B(X, Y) - B(Y, X)].$$

Actually, we can express A in terms of the torsion tensors T and $\bar{T}$ of D and $\bar{D}$, respectively, for $2A(X, Y) =$

$$\bar{D}_X Y - D_X Y - \bar{D}_Y X + D_Y X = \bar{T}(X, Y) + [X, Y] - T(X, Y) - [X, Y] = \bar{T}(X, Y) - T(X, Y)$$

THEOREM. *The following statements are equivalent:*

(a) *The connections D and $\bar{D}$ have the same geodesics.*

(b) $B(X, X) = 0$ *for all vectors X.*

(c) $S = 0$.

(d) $B = A$.

Proof. (a) implies (b): Take X at m in M and let g be the geodesic with initial vector X. Extend X along g by letting X be the tangent to g; then $B(X, X) = \bar{D}_X X - D_X X = 0 - 0$, since g is a geodesic for both connections.

(b) implies (a): Let g be a geodesic for D with tangent field X; then $\bar{D}_X X = B(X, X) + D_X X = 0$ on g; hence g ie a geodesic for $\bar{D}$.

(b) equivalent to (c): Since S is symmetric, it is determined by its diagonal values $S(X, X)$, and $B(X, X) = 0$ iff $S(X, X) = 0$.

(c) equivalent to (d): For $B = S + A$.//

THEOREM. *The connexions D and $\bar{D}$ are equal iff they have the same geodesics and the same torsion tensors.*

Proof. That the first part implies the second is trivial. Conversely, if the geodesics are the same, then $S = 0$, and if the torsion tensors are equal, then $A = 0$; hence $B = 0$ and $D = \bar{D}$.//

THEOREM. *Given a connexion $\bar{D}$ on M, there is a unique connexion D having the same geodesics as $\bar{D}$ and zero torsion.*

Proof. Let $D_X Y = \bar{D}_X Y - (1/2)\bar{T}(X, Y)$. It is trivial to check that D satisfies the required properties to define a connexion. Here $B = (1/2)\bar{T} = A$, since a torsion tensor is skew-symmetric; thus $S = 0$, so D and $\bar{D}$ have the same geodesics. Moreover, $T = \bar{T} - 2A = 0$, so D has zero torsion. The uniqueness follows from the preceding theorem.//

Thus if we partition connexions into equivalence classes by placing two connexions with the same geodesics in the same class, then in each class there exists a unique torsion-free (zero torsion) connexion. Moreover, given any connexion D and any skew-symmetric vector-valued 2-covariant tensor $\bar{T}$, there exists a connexion with torsion tensor $\bar{T}$ and the same geodesics as D. From the above proof we have $\bar{T}(X, Y) = 2(\bar{D}_X Y - D_X Y)$, which provides a geometric interpretation of the torsion tensor of a connexion as measuring the difference between covariant differentiation in the given connexion and covariant differentiation in the torsion-free connexion with the same geodesics.

Section 5.5. Bundle viewpoint.

In this section we define a connexion on the bundle of bases over a manifold and sketch a proof of the equivalence of such a definition with our previous viewpoints. This is the fourth (and last) viewpoint we consider. The bundle viewpoint provides a natural "jumping off" for generalizations to connexions in all kinds of bundles, and much of the research in differential geometry at this time uses these concepts. For more details the reader is referred to the book by Crittender and Bishop or the book by Kobayashi and Nomizu.

Throughout this section let M be a C^∞ n-manifold, let $B = B(M)$ be the bundle of bases over M (see problem 22), and let $\pi\colon B \to M$ be the natural projection map. If D is a connexion on M, then by integrating ordinary differential equations ((5) above), we can parallel translate the tangent space along curves in M. If $b = (m; e_1, \ldots, e_n)$ is in B and σ is a curve in M with $\sigma(0) = m$, then by parallel translation we define a C^∞ curve $\bar{\sigma}(t) = (\sigma(t); e_1(t), \ldots, e_n(t))$ in B, where $e_i(t)$ is the parallel translate of $e_i = e_i(0)$ along σ to $\sigma(t)$. Since $\pi \circ \bar{\sigma} = \sigma$, we say $\bar{\sigma}$ is a "lift of σ", or $\bar{\sigma}$ "lies over σ," and since $\bar{\sigma}$ reads off a parallel base, we say $\bar{\sigma}$ is a "horizontal" curve in B. Thus a connexion D on M yields unique "horizontal lifts" of C^∞ curves in M. The bundle definition of a connexion gives an independent method for defining "horizontal lifts" (of curves in M) with the correct properties.

Recall at each point b in B we defined the subspace of *vertical vectors* $V_b = [X \text{ in } B_b\colon\ \pi_*(X) = 0]$. A *connexion on* B is a mapping H that assigns to each b in B subspace H_b of B_b such that:

(1) $H_b \cap V_b = 0$ and $\pi_*|H_b$ is an isomorphism of H_b onto $M_{\pi(b)}$ (hence H_b is n-dimensional).

(2) $(R_g)_*(H_b) = H_{bg}$ for all g in $GL(n, R)$.

(3) H is C^∞; i.e., for each b in B there is a neighborhood U and a set of n independent C^∞ vector fields $E_1, \ldots, E_n$ on U that give a base for $H_{b'}$ for every b' in U.

If X is in H_b, we say X is a *horizontal vector.* Property (1) implies for each X in B_b there is a unique decomposition $X = X_H + X_V$ with X_H in H_b and X_V in V_b, and property (3) implies if X is C^∞ then X_H and X_V are C^∞ fields. If X is a C^∞ field with domain U in M, then there is a *unique C^∞ horizontal field* $\bar{X}$ on $\bar{U} = \pi^{-1}(U)$ with $\pi_*(\bar{X}_b) = X_{\pi(b)}$ for all b in $\bar{U}$.

Having the existence of "horizontal lifts" for vector fields, one can "horizontally" lift curves in a natural way. Thus if σ is a curve in M with tangent T (non-vanishing), extend T to a C^∞ field in a neighborhood U of a univalent part of σ, lift T to a horizontal field $\bar{T}$ on $\bar{U}$, and take integral curves of $\bar{T}$ to find horizontal lifts of σ. The parallel translation so defined will be independent of the base (the starting point for $\bar{\sigma}$) by property (2); i.e., if $\bar{\sigma}$ is horizontal (has a horizontal tangent), then $R_g \circ \bar{\sigma}$ is also horizontal.

There is a dual viewpoint involving differential forms. To motivate it, let H be a connexion as described above and notice at each $b = (m; e_1,\dots,e_n)$ in B we can define a unique horizontal field $E_i(b)$ with $\pi_*(E_i(b)) = e_i$ by (1). The fields $E_1,\dots,E_n$ are global independent horizontal C^∞ fields on B. Together with the natural vertical fields $E_{11},\dots,E_{nn}$, we get a global base field on B. Let $\overline{w}_1,\dots,\overline{w}_n$, $\overline{w}_{11},\dots,\overline{w}_{nn}$ be the dual 1-forms to this base (where $\overline{w}_1,\dots,\overline{w}_n$ are the natural 1-forms of problem 41). Then if X in B_b, $X_V = \Sigma^n_{i,j=1}\overline{w}_{ij}(X)(E_{ij})_b$. If one knows X_V, then, of course, $X_H = X - X_V$. Thus giving X_H (or giving H) is equivalent to giving "vertical projections" at each point in B. Thus a set of *connection 1-forms* $\overline{w}_{ij}$ (for i, $j = 1,\dots,n$) on B is a set of 1-forms such that

(1') $\overline{w}_{ij}|V_b$ form a dual base to E_{ij} at all b in B,

(2') $\overline{w}_{ij}((R_g)_*X) = \Sigma^n_{r,s=1}g^{-1}_{ir}\overline{w}_{rs}(X)g_{sj}$ for all X in B_b,

(3') $\overline{w}_{ij}$ are C^∞ for all i and j.

That the definition of a connexion on B in terms of H or in terms of $\overline{w}_{ij}$ is equivalent is left as a problem.

Notice that the $\overline{w}_{ij}$ can be used to define a *Lie algebra* (of $GL(n, R)$) *valued* 1-*form* $\overline{w}$ by $\overline{w}(X) = \Sigma^n_{i,j=1}\overline{w}_{ij}(X)X_{ij}$, where the X_{ij} are the canonical left invariant fields on $GL(n, R)$ (see problem 21).

Finally, we connect with the Cartan viewpoint. Let $e_1,\dots,e_n$ be a base field on the open set U in M. Define a C^∞ map $f\colon U \to B$ by $f(m) = (m; (e_1)_m,\dots,(e_n)_m)$ for m in U. Since $\pi \circ f$ is the identity of U, we call f a *section* over U. Let w_{ij} be the connexion forms defined in section 5.2, and let $\overline{w}_{ij}$ be the global forms defined above. Then $w_{ij} = \overline{w}_{ij} \circ f_*$ on U.

Thus the Cartan structural equations (13) and (14) (and the torsion and curvature 2-forms) can be carried up to global equations on B.

Problems. Let M and M' be C^∞ manifolds.

42. Let $x_1 = x$ and $x_2 = y$ be the usual coordinates on R^2. Define a connexion D on R^2 by letting $\Gamma^i_{jk} \equiv 0$ except for $\Gamma^1_{12} = \Gamma^1_{21} \equiv 1$. Set up and solve the differential equations for the geodesics thru any point in R^2. Find the particular geodesic g with $g(0) =$

(2, 1) and $T_g(0) = (\partial/\partial x) + (\partial/\partial y)$. Is D complete? Do the geodesics emanating from the origin pass thru all points of the plane? If σ and γ are geodesics with $\gamma(0) = \sigma(0)$, and $T_\gamma(0) = bt_\sigma(0)$ for b in R, show $\gamma(t) = \sigma(bt)$ for all possible t. Investigate the connexion D' with all $\Gamma'^i_{jk} = 0$ except $\Gamma'^1_{12} = 1$.

43. Let D be a connexion on M. Let $\sigma(t)$ be an integral curve of the C^∞ field X, let $e_1(t),\dots,e_n(t)$ be a parallel base along σ, and let $Y(t) = \Sigma y_i(t)e_i(t)$ be a C^∞ field along σ. Show $(D_XY)(t) = \Sigma(dy_i/dt)e_i(t)$ along σ. Show $(D_XY)(0) = \lim (1/t)[(P_{t,0}Y(\sigma(t)) - Y(0)]$ as $t \to 0$.

44. Let f be a connexion preserving C^∞ map of M into M'. Show $f_*(\mathrm{Tor}\,(X, Y)) = \mathrm{Tor}'\,(f_*X, f_*Y)$ and $f_*(R(X, Y)Z) = R'(f_*X, f_*Y)(f_*Z)$.

45. A manifold M is *parallelizable* if there is a connexion D on M in which parallel translation is independent of curves, and such a D is called a *flat connexion*. Show M is parallelizable iff there is a global C^∞ base field on M. If D is a flat connexion, show its curvature tensor is zero (see problem 85).

46. Let G be a Lie group. Define the *left invariant connexion D* on G by asserting all vector fields in the Lie algebra $\mathcal{G}$ are parallel fields. Show D is flat, G is parallelizable, and if X and Y are in $\mathcal{G}$, then $\mathrm{Tor}\,(X, Y) = -[X, Y]$. Show that each geodesic g on G is the left translate of a one-parameter subgroup σ; i.e., $g(t) = L_{g(0)}(\sigma(t))$ for all t. Show D is complete.

47. Let D be a connexion on M. For m in M, let H_m denote the set of linear maps of M_m into itself, obtained by parallel translation of M_m around broken C^∞ curves starting and ending at m. Show H_m is a group. If M is connected, show H_m is isomorphic to $H_{m'}$ for m' in M. The group H_m is called the *holonomy group* at m, and if M is connected, then the *holonomy group of M* is the group $H = H_m$ for any m in M. Restricting the closed curves to be null-homotopic, one obtains the *restricted holonomy group* H^0_m. If D is flat, show $H_m = 0$. If M is the unit sphere in R^3 and D is the Riemannian connexion, show that $H = SO(2, R)$, where $SO(2, R)$ is the *special orthog-*

onal group, or *rotation group*, consisting of orthogonal maps with determinant one.

48. (Continuing problem 13.) Let $X = \partial/\partial x_1$ and $Y = \partial/\partial x_2$ for a coordinate system $x_1, \ldots, x_n$ on M about m, and show $\lim_{t \to 0} (1/t)[(P_{0,t} - I)(\partial/\partial x_j)] = R(Y, X)(\partial/\partial x_j) = \Sigma_k R^k_{j\,2\,1}(\partial/\partial x_k)$, where I is the identity map and $P_{0,t}$ is parallel translation along γ from $\gamma(0)$ to $\gamma(t)$. Because of this, one often says $R(X, Y)$ is "infinitesional parallel translation around an infinitesimal parallelogram spanned by X and Y."

6. Riemannian Manifolds and Submanifolds

The definition of a Riemannian (and a semi-Riemannian) manifold was given in section 2.1. A manifold on which one has singled out a specific symmetric and positive definite (or non-singular) 2-covariant tensor field, called the metric tensor, is a Riemannian (or semi-Riemannian) manifold. In this chapter we generalize the theory of Chapters 2 and 3 in a natural way. Much of the theory applies to semi-Riemannian manifolds and submanifolds, but, in general, we phrase things only in Riemannian terms.

Section 6.1. *Length and distance.*

The metric tensor allows us to define lengths, angles, and distances. Let M be a Riemannian manifold with metric tensor $<,>$. Let X and Y be in M_m. Define the *length* of X by $|X| = \sqrt{<X, X>}$. Define the *angle θ between X and Y* (both non-zero) by letting $<X, Y> = |X||Y| \cos \theta$ where $0 \leq \theta \leq \pi$, and notice the *Schwartz inequality* $|<X, Y>| \leq |X||Y|$ makes this possible.

The length of a curve is now defined by integrating the length of its tangent vector field. Let σ be a C^∞ curve on $[a, b]$ with tangent field T (or T_σ if necessary). The *length of σ from a to be*, denoted by $|\sigma|_a^b$, is defined by

(1) $$|\sigma|_a^b = \int_a^b \sqrt{<T(t), T(t)>}dt.$$

The integral exists, since the integrand is continuous. The length of a broken C^∞ curve is defined as the (finite) sum of the lengths of its C^∞ pieces. The number $|\sigma|_a^b$ is independent of the parameterization of its image set in the following sense: let g be a C^1 map of $[c, d]$ into $[a, b]$ with end points mapping to end points (assume $g(c) = a$ and $g(d) = b$); then

$$\int_a^b \langle T_\sigma(t), T_\sigma(t)\rangle^{1/2} dt = \int_c^d \langle T_\sigma(g(t)), T_\sigma(g(t))\rangle^{1/2} g'(t)dt$$

$$= \int_c^d \langle T_{\sigma\circ g}(t), T_{\sigma\circ g}(t)\rangle^{1/2} dt$$

since $T_{\sigma\circ g}(t) = g'(t)T_\sigma(g(t))$ by the chain rule. Thus we can write $|\sigma|_q^p = |\sigma|_a^b$, where $q = \sigma(a)$ and $p = \sigma(b)$.

Classically, the metric tensor is almost always expressed by the notation "$ds^2 = g_{ij}dx_idx_j$." This means one is giving the inner product on a coordinate domain U with coordinate functions $x_1,\ldots,x_n$ in terms of the coordinate bases; i.e., if $X_i = \partial/\partial x_i$, then $g_{ij} = \langle X_i, X_j\rangle$ is a C^∞ function on U. If $Y = \Sigma y_iX_i$ and $Z = \Sigma z_kX_k$, then $\langle Y, Z\rangle = \Sigma_{i,k=1}^n y_iz_kg_{ik}$. Thus, giving the matrix of functions g_{ij} on U determines the inner product on U. The "ds" only makes sense when one is discussing a curve σ which maps into U, for then let $s(t) = |\sigma|_a^t$ and

$$\left(\frac{ds}{dt}\right)^2 = \langle T, T\rangle = \Sigma g_{ij} \frac{d(x_i \circ \sigma)}{dt} \frac{d(x_j \circ \sigma)}{dt} .$$

If M is connected, a *pseudo-metric* is defined on M by

(2) $\qquad d(p, m) = \inf\ [|\sigma|:\ \sigma$ a broken C^∞ curve from p to $m]$.

Trivially, $d(p, m) \geq 0$, $d(p, p) = 0$, and $d(p, m) = d(m, p)$. The triangle inequality is left as a problem.

THEOREM. *The pseudo-metric topology on M equals the manifold topology.*

Proof. (After Seifert and Threlfall, p. 44). Let m be any point in M, and let $x_1,\ldots,x_n$ be a coordinate system about m with domain U.

For p in U let $d(p) = d(m, p)$ defined above, and let $d'(p) = [\Sigma x_i(p)^2]^{1/2}$ where we assume $x_i(m) = 0$. Choose $a > 0$ so $A = [p\colon\ d'(p) \leq a]$ is contained in U. On the compact set $B = [(p, X_p)\colon\ p$ in A and $1 = \Sigma dx_i(X_p)^2]$, the norm function, $|X_p| = [\Sigma_{ij} g_{ij}(p) dx_i(X_p) dx_j(X_p)]^{1/2}$, is a continuous function which takes on a maximum R and a minimum $r > 0$.

Let σ be any broken C^∞ curve in A with $\sigma(0) = m$, $\sigma(b) = p$ and $(\sigma(t), T_\sigma(t))$ always in B. Then $|\sigma| = \int_0^b |T_\sigma(t)| dt \geq rb \geq rd'(p)$. For a broken curve σ from m to p that leaves A, one has $|\sigma| \geq ra \geq rd'(p)$. Hence, (1) $d(p) \geq rd'(p)$. But if σ curve with $x_i \circ \sigma(t) = tp_i/d'(p)$, where $x_i(p) = p_i$, then $|\sigma| = \int_0^{d'(p)} |T_\sigma(t)| dt \leq Rd'(p)$. Hence, (2) $d(p) \leq Rd'(p)$.

The inequalities (1) and (2) prove the theorem.//

Corollary. A connected Riemannian manifold M is Hausdorff iff the pseudo-metric d is a metric.

In Chapter 10 we show that geodesics are the curves that locally minimize arc length, i.e., the length of a small piece of a geodesic in M is precisely the distance between the end points of the piece.

Henceforth we assume all manifolds we mention are Hausdorff. A Riemannian manifold is *complete* if it is complete as a metric space, i.e., every Cauchy sequence must converge.

Section 6.2. Riemannian connexion and curvature.

A *Riemannian connexion* D on a Riemannian manifold M is a connexion D such that

(3) $\quad D_X Y - D_Y X = [X, Y]$, and

(4) $\quad Z\langle X, Y\rangle = \langle D_Z X, Y\rangle + \langle X, D_Z Y\rangle$,

for all fields X, Y, and Z with a common domain. The fundamental theorem of (semi-) Riemannian manifolds is the following:

THEOREM. *There exists a unique Riemannian connection on a (semi-) Riemannian manifold.*

Proof. We show a Riemannian connexion D exists and is unique on every coordinate domain U. The uniqueness implies D must agree on overlapping domains; hence D exists and is unique on all of M.

Let $X_1, \ldots, X_n$ be the coordinate fields on U, let $g_{ij} = \langle X_i, X_j \rangle$ on U, and let $(g^{-1})_{ij}$ be the jth entry of the inverse matrix of $g = (g_{ij})$ (which is non-singular). If (3) and (4) hold, then

$$(5) \qquad X_i\langle X_r, X_j\rangle + X_j\langle X_r, X_i\rangle - X_r\langle X_i, X_j\rangle = 2\langle D_{X_i}X_j, X_r\rangle$$

since $[X_k, X_s] = 0$ for all k, s. By section 5.2, giving D on U is equivalent to giving functions Γ^i_{jk} with $D_{X_k}(X_j) = \Sigma^n_{i=1}\Gamma^i_{jk}X_i$ and demanding properties (1) through (4) of section 5.1 are valid. Thus (5) implies $2\Sigma_k\Gamma^k_{ji}g_{kr} = X_ig_{rj} + X_jg_{ri} - X_rg_{ij}$; hence

$$(6) \qquad \Gamma^k_{ij} = 1/2\Sigma_r(g^{-1})_{kr}\left(\frac{\partial g_{rj}}{\partial x_i} + \frac{\partial g_{ri}}{\partial x_j} - \frac{\partial g_{ij}}{\partial x_r}\right).$$

This is the classical expression for the Christoffel function Γ^k_{ij} in terms of the metric tensor. Use (6) to define D on U. A direct check of (3) and (4) shows D is Riemannian, and the explicit representation (6) shows D is unique.//

The above theorem is special case of a more general theorem (problem 70). For the rest of this section let M be a (semi) Riemannian manifold and let D be the Riemannian connexion on M. The *Riemann-Christoffel curvature tensor* (of type 0, 4) is the 4-covariant tensor $K(X, Y, Z, W) = \langle X, R(Z, W)Y\rangle$ for X, Y, Z, and W in M_m.

THEOREM. *The following relations are true:*

(a) $R(X, Y)Z + R(Z, X)Y + R(Y, Z)X = 0$

(b) $K(X, Y, Z, W) = -K(Y, X, Z, W)$

(c) $K(X, Y, Z, W) = -K(X, Y, W, Z)$

(d) $K(X, Y, Z, W) = K(Z, W, X, Y)$.

The relation (a) is called the *first Biachi identity* and it holds for any symmetric connexion. These relations are equivalent to the "symmetries" of the indices of the classical R_{ijkh} functions.

Proof. For (a), use the Jacobi identity, property (3) above, and compute. For (c), use $R(Z, W) = -R(W, Z)$. For (b), use property (4) to shift D from one slot to the other. For (d), notice (a) implies (a'): $K(X, Y, Z, W) + K(X, W, Y, Z) + K(X, Z, W, Y) = 0$. By writing (a') three more times, cyclicly permuting the arguments of the first term one step from one line to the next, adding all four equations, and cancelling via (b) and (c), one obtains (d).//

For X and Y in M_m, let

$$(7) \qquad A(X, Y) = \langle X, X\rangle\langle Y, Y\rangle - \langle X, Y\rangle^2.$$

If $A(X, Y) \neq 0$, let

$$(8) \qquad \overline{K}(X, Y) = K(X, Y, X, Y)/A(X, Y),$$

and by direct computations, using the above properties of K, one can show

$$\overline{K}(X, Y) = \overline{K}(Y, X) = \overline{K}(rX, sY) = \overline{K}(X + tY, Y)$$

for r, s, and t not zero. Thus if $A(X, Y) \neq 0$ and $ad - bc \neq 0$, then $\overline{K}(X, Y) = \overline{K}(aX + bY, cX + dY)$, and we define $K(P)$, the *Riemannian curvature of the 2-dimensional subspace P of M_m spanned by X and Y, by* $K(P) = \langle X, R(X, Y)Y\rangle/A(X, Y)$. In section 2.4, we showed $K(M_m) = K(m)$ is the Gauss curvature of a surface M in R^3. In the Riemannian case, $[A(X, Y)]^{1/2}$ is the *area of the parallelogram spanned by X and Y.*

Let $f: M \to M'$ be a C^∞ map between Riemannian manifolds. If there is a C^∞ real valued positive function F on M such that for all m in M and all X, Y in M_m, $\langle f_*X, f_*Y\rangle = F(m)\langle X, Y\rangle$, then f is a *conformal* (or *strictly conformal*) map and F is called the *scale function.* If F exists but $F \geq 0$ only, then f is *weakly conformal.* If $F = 1$, then f is *an isometry.* If f is an isometry and a diffeomorphism, then f is *isometric* and M is *isometric* to M'. If F is constant, then f is *homothetic.*

At this point, we explicitly call the reader's attention to problem 52, which is considered an integral part of the theory of Riemannian manifolds.

Section 6.3. Curves in Riemannian manifolds.

This section parallels the standard treatment of curves in advanced calculus. Let M be a Riemannian manifold with Riemannian connexion D. Let σ be a C^∞ curve in M with tangent field $V = \sigma_*(d/dt)$, which can legitemately be called the "velocity vector" of σ since "length" is defined. Assuming V does not vanish on the domain of σ, define the *unit tangent vector* $T(t) = V(t)/|V(t)|$, and define the *speed* function $s' = (ds/dt) = |V(t)|$, so $V(t) = s'(t)T(t)$ for t in the domain of σ. Define the *geodesic curvature vector field* of σ to be the field $D_T T$, and its length k_1 is the *geodesic curvature* of σ. Notice that $D_T T$ and k_1, at a particular point on the curve, do not depend on the parameterization of the "point set of the curve" but only on the orientation (choice of "direction") and the existence of a C^∞ parameterization with non-vanishing tangent at the point.

The curve σ is a geodesic ($D_V V = 0$) iff V has constant length and (a) $D_T T = 0$ or (b) $k_1 = 0$. This follows since $D_V V = s'D_T(s'T) = s's''T + (s')^2 D_T T$ and $s' > 0$ while $D_T T$ is orthogonal to T, ($\langle T, T\rangle = 1$ so $0 = T\langle T, T\rangle = 2\langle D_T T, T\rangle$).

When $k_1(t) > 0$, define the (first) *normal* to σ at $\sigma(t)$ to be the unit vector $N_1(t)$ such that $D_T T = k_1 N_1$ at t. If N_1 is defined on an interval, then, $0 = T\langle N_1, T\rangle = \langle D_T N_1, T\rangle + \langle N_1, D_T T\rangle = \langle D_T N_1, T\rangle + k_1$, so $D_T N_1 \neq 0$ on the interval. The vector $D_T N_1 + k_1 T$ is orthogonal to both T and N_1; hence, let its length be k_2, the *second curvature* or *torsion*. If $k_2(t) > 0$, define the *second normal* to σ at $\sigma(t)$ to be the unit vector $N_2(t)$ such that $D_T N_1 + k_1 T = k_2 N_2$. If $k_2 > 0$ on an interval, then the above process can be continued to define k_3, and where $k_3 > 0$, one gets N_3, etc. The vectors $T, N_1, N_2, \ldots$ are called *Frenet vectors*, and the equations that express the $D_T N_i$ in terms of the Frenet vectors are called the *Frenet formulae*.

When M is a 2-manifold and $k_1 > 0$, then the Frenet formulae become $D_T T = k_1 N$ and $D_T N_1 = -k_1 T$. In this case it is possible to locally choose N_1 along σ independently of $D_T T$ (on univalent pieces of σ), and letting $D_T T = k_1 N_1$ would define k_1, which could take on negative values (see problem 72).

Section 6.4. Submanifolds.

The theory in sections 2.3 and 2.4 is now generalized. Throughout this section let the k-manifold M be a (non-singular) submanifold of the (semi-) Riemannian manifold $\bar{M}$. In the semi-Riemannian case, the submanifold M is *non-singular* if the metric tensor is non-singular when restricted to M_m for all m in M (thus M is a semi-Riemannian manifold under the induced metric tensor). The induced metric tensor on M is called the *first fundamental form* on M. Let $\bar{D}$ be the Riemannian connexion on $\bar{M}$.

THEOREM. *For C^∞ fields X and Y with domain A on M (and tangent to M), define D_XY and $V(X, Y)$ on A by decomposing $\bar{D}_XY$ into its unique tangential and normal components, respectively; thus,*

$$\bar{D}_XY = D_XY + V(X, Y). \tag{9}$$

Then D is the Riemannian connexion on M and V is a symmetric vector-valued 2-covariant C^∞ tensor called the second fundamental tensor. *The decomposition equation* (9) *is called the* Gauss equation.

Proof. We will establish the C^∞ nature of the decomposition. The rest of the proof will only be outlined, for it is a simple exercise. Use the properties of $\bar{D}$ (since it is a connexion) to establish the properties of D (making it a connexion) and the tensor character of V (its multilinearity). Zero torsion for $\bar{D}$ implies zero torsion for D, and V is symmetric (use the proposition in section 2.2, which generalizes trivially). Since $\bar{D}$ satisfies condition (4) (section 6.2), D does too. Hence D is Riemannian, and by the uniqueness theorem, D is the Riemannian connexion on M.

To show D and V are C^∞ on A, choose p in A. Let $\bar{U}$ and U be special coordinate domains about p in $\bar{M}$ and M, respectively, with $U \subset A$, and let $\bar{Z}_1,\ldots,\bar{Z}_n$ and $Z_1 = \bar{Z}_1|_U,\ldots,Z_k = \bar{Z}_k|_U$ be the coordinate vector fields on $\bar{U}$ and U, respectively. Apply the Gram-Schmidt process to $\bar{Z}_1,\ldots,\bar{Z}_n$ on $\bar{U}$ to obtain C^∞ (the Gram-Schmidt process is algebraic) orthonormal fields $W_1,\ldots,W_n$ on $\bar{U}$ such that $W_1|_U,\ldots,W_k|_U$ give a C^∞ orthonormal base of M_m for m in U, while $W_{k+1}|_U,\ldots,W_n|_U$ give $\bar{M}$-vector fields that are C^∞ on U and form a base of the orthogonal complement to M_m, for m in U. Let $X = \Sigma_{i=1}^k x_iW_i$ and $Y = \Sigma_{j=1}^k y_jW_j$ define C^∞ functions x_i and y_i on U for $i = 1,\ldots,k$, and let

$\overline{D}_{W_i} W_j = \Sigma_{r=1}^{n} B_{ji}^{r} W_r$ define C^∞ functions B_{ij}^{r} on $\overline{U}$. Then

$$\overline{D}_X Y = \Sigma(XY_j)W_j + \Sigma y_j B_{ji}^{r} W_r$$

where i and $j = 1,\ldots,k$ and $r = 1,\ldots,n$; thus

$$D_X Y = \Sigma_{r=1}^{k}[(Xy_r) + \Sigma_{i,j=1}^{k} y_j x_i B_{ji}^{r}]W_r$$

and

$$V(X, Y) = \Sigma_{r=k+1}^{n}[\Sigma_{i,j=1}^{k} y_j x_i B_{ji}^{r}]W_r$$

are C^∞ on U.//

By decomposing the curvature $\overline{R}$ into tangent and normal parts, we obtain the *Gauss curvature equation* (10), and the *Codazzi-Mainardi equation* (11), respectively. Let X, Y, and Z be C^∞ fields tangent to M with a common domain. Writing the decomposition of a vector W as $W = \tan W + \text{nor } W$,

$$\text{(10)} \qquad \tan \overline{R}(X, Y)Z = R(X, Y)Z + \tan [\overline{D}_X V(Y, Z) - \overline{D}_Y V(X, Z)],$$

and

$$\text{(11)} \qquad \text{nor } \overline{R}(X, Y)Z = V(X, D_Y Z) - V(Y, D_X Z) - V([X, Y], Z) + \text{nor } [\overline{D}_X V(Y, Z) - \overline{D}_Y V(X, Z)].$$

Since V is a tensor, i.e., $V(X_m, Y_m)$ is well-defined and independent of the fields X and Y used to compute it in the Gauss equation, we define X and Y to be *conjugate vectors* at m if $V(X, Y) = 0$. A vector X in M_m is an *asymptotic vector* if $V(X, X) = 0$, and in any case, define the *asymptotic* (or *normal*) *curvature* of X, k_X, by $k_X = |V(X, X)|$. If $V_m = 0$, then m is a *flat point* on M.

If σ is a curve in M with C^∞ unit tangent T, then $V(T, T)$ is the *normal curvature vector field* of σ and $k_T = |V(T, T)|$ is the *normal curvature* of σ.

THEOREM (Meusnier). *All curves on M with the same unit tangent T at a point have the same normal curvature at that point. If σ a curve on M with C^∞ unit tangent T, then $(\bar{k}_1)^2 = (k_1)^2 + (k_T)^2$ relates the geodesic curvatures $\bar{k}_1$ and k_1 of σ in $\bar{M}$ and M with its normal curvature k_T. Moreover, $k_T = \bar{k}_1 \cos\phi$ determines the angle ϕ between the normal $\bar{N}_1$ of σ in $\bar{M}$ and the normal curvature vector $V(T, T)$ if ϕ is defined.*

Proof. The first sentence follows since V is a tensor. The second sentence follows from the Gauss equation $\bar{D}_T T = D_T T + V(T, T)$ since the vectors on the right are orthogonal. For the third sentence, if $\bar{k}_1 = 0$, then $k_1 = k_T = 0$ and ϕ not defined. If $\bar{k}_1 > 0$ and $k_T = 0$, then $V(T, T) = 0$, $\bar{N}_1$ is tangent to M, and $\phi = \pi/2$ (if anything). If $k_T \neq 0$, let N be the unit normal in direction of $V(T, T)$ and

$$k_T = \langle V(T, T), N\rangle = \langle \bar{D}_T T, N\rangle = \bar{k}_1 \cos\phi.//$$

The theorem and corollary at the end of section 2.3 can now be generalized by replacing R^n by $\bar{M}$.

Section 6.5. Hypersurfaces.

In this section, let M be a hypersurface in the Riemannian manifold $\bar{M}$ and let N be a C^∞ unit normal on M. Define the *Weingarten map* $L(X) = \bar{D}_X N$ for X in M_m (as in section 2.2). The Gauss equation for M now becomes

$$(12) \qquad \bar{D}_X Y = D_X Y - \langle LX, Y\rangle N$$

since $\langle V(X, Y), N\rangle = \langle \bar{D}_X Y, N\rangle = X\langle N, Y\rangle - \langle Y, L(X)\rangle$ and $\langle N, Y\rangle \equiv 0$. Thus $V(X, Y) = -\langle LX, Y\rangle N$.

The fundamental forms and the imbedded geometric invariants of M in $\bar{M}$ are defined in terms of L exactly as in section 2.2. Notice in this case V is symmetric is equivalent to L being self adjoint.

The Gauss curvature equation (10) and Codazzi-Mainardi equation (11) now become

$$(13) \qquad \tan \bar{R}(X, Y)Z = R(X, Y)Z - [\langle LY, Z\rangle L(X) - \langle LX, Z\rangle L(Y)],$$

and

(14) $\quad \text{nor}\, \bar{R}(X, Y)Z = -\langle D_X LY - D_Y LX - L[X, Y], Z\rangle N$

respectively.

The torsion tensor is generalized by defining for any C^∞ linear transformation valued tensor $W_p\colon M_p \to M_p$, on a C^∞ manifold M, the *torsion of* W, Tor_W, by

(15) $\quad \text{Tor}_W(X, Y) = D_X W(Y) - D_Y W(X) - W[X, Y].$

The Codazzi-Mainardi equation (14) on a hypersurface becomes $\text{nor}\, \bar{R}(X, Y)Z = -\langle \text{Tor}_L(X, Y), Z\rangle N$. Thus $\text{Tor}_L = 0$ on M iff

(16) $\quad \bar{R}(X, Y)Z = R(X, Y)Z - [\langle LY, Z\rangle LX - \langle LX, Z\rangle LY].$

The following theorem generalizes the "theorema egregium" of Gauss, and actually, it may be generalized to the case where M is a k-submanifold of $\bar{M}$ (see Hicks [1]).

THEOREM. *Let M be a hypersurface in the Riemannian $\bar{M}$, let P be a 2-dimensional subspace of M_m, and let $K(P)$ and $\bar{K}(P)$ be the Riemannian curvature of P in M and $\bar{M}$ respectively. Let N be a unit C^∞ normal on a neighborhood of m, and let $L(X) = \bar{D}_X N$ for X in M_M. If X and Y form an orthonormal base of P, then*

(17) $\quad \bar{K}(P) = K(P) - (\langle LY, Y\rangle\langle LX, X\rangle - \langle LX, Y\rangle^2).$

Proof. Combine the definition of Riemannian curvature with the Gauss curvature equation (13).//

When $\bar{M}$ is a 3-manifold, the above theorem shows the determinant of L is independent of the imbedding (i.e., independent of L) but depends only on the Riemannian structure of $\bar{M}$ and M.

A related result is a form of the *Lemma of Synge*.

THEOREM. *Let $k > 1$, and let M be a k-submanifold of the Riemannian n-manifold $\bar{M}$. Let g be a geodesic of $\bar{M}$ that lies in M, let T be the unit tangent to g, let X be a unit field tangent to M which* *is parallel in M along g and orthogonal to T, and let P be the subspace spanned by X and T. Then $\bar{K}(P) \geq K(P)$ along g, and $\bar{K}(P) = K(P)$ iff X is parallel along g in $\bar{M}$.*

Proof. We prove the theorem for $k = n - 1$, leaving the other cases to problem 55. Let N be a C^∞ unit normal on a neighborhood of a point on g and let $L(Z) = \bar{D}_Z N$. Here $\bar{D}_T T = 0$ so $D_T T = 0$ and $\langle LT, T\rangle = 0$. By the previous theorem, $\bar{K}(P) = K(P) + \langle LX, T\rangle^2 \geq K(P)$. If equality holds, then $\langle LX, T\rangle = 0$ so $\bar{D}_T X = D_T X = 0$, and conversely.//

There is a basic "rigidity" theorem for hypersurfaces of R^n which is our next goal. This theorem is a uniqueness theorem, and there is a corresponding existence theorem that is proved in Chapter 9. When $n = 3$, the theorem was first proved by 0. Bonnet (1867).

Intuitively, this theorem states if two hypersurfaces in R^n are isometric and their normals are "bending the same", then by a "rigid motion" one can superimpose the two manifolds.

THEOREM. *Let M and M' be connected hypersurfaces in R^n for $n \geq 3$. Let N and N' be C^∞ unit normal fields on M and M', respectively. Let F be a diffeomorphism of M onto M' that preserves the first and second fundamental forms. Then there is an isometry G of R^n with $F = G|_M$.*

Proof. During this proof let us use "primes" to denote concepts belonging M' which correspond to familiar concepts for M; i.e., let $L(X) = \bar{D}_X N$ for X in M_p and $L'(Y) = \bar{D}_Y N'$ for Y in $M'_{p'}$. The hypothesis states if X and Z are in M_p, then

$$\langle F_*X, F_*Z\rangle = \langle X, Z\rangle \text{ and } \langle L'(F_*X), F_*Z\rangle = \langle LX, Z\rangle.$$

Combining these statements,

$$\langle L'(F_*X), F_*Z\rangle = \langle LX, Z\rangle = \langle F_*LX, F_*Z\rangle$$

for all Z which implies $L' \circ F_* = F_* \circ L$. Thus the hypothesis could be rephrased as a demand that F be an isometry of M onto M' whose Jacobian commutes with the Weingarten maps. Since an isometry is connexion preserving, $F_*(D_X Z) = D'_{F_* X} F_* Z$ for vectors X and fields Z tangent to M.

If p in M, we extend the Jacobian of F to be a linear map of $(R^n)_p$ onto $(R^n)_{p'}$ where $p' = F(p)$. Let W be in $(R^n)_p$, then $W = W_t + aN_p$ where W_t is tangent to M, so define

$$F_*(W) = F_*(W_t) + aN'_{p'}.$$

If X is in M_p and W is a C^∞ field of R^n-vectors on M, then

$$F_*(\overline{D}_X W) = \overline{D}_{F_* X}(F_* W),$$

where $\overline{D}$ is a natural covariant differentiation on R^n. This follows since

$$\overline{D}_X W = \overline{D}_X W_t + \overline{D}_X(aN)$$

$$= D_X W_t - \langle LX, W_t\rangle N + (Xa)N + aLX,$$

and

$$F_*(\overline{D}_X W) = D'_{F_* X} F_* W_t - \langle F_* LX, F_* W_t\rangle N' + F_* X(a \circ F^{-1})N' + (a \circ F^{-1})L'F_*$$

$$= \overline{D}_{F_* X} F_* W_t + \overline{D}_{F_* X}((a \circ F^{-1})N') = \overline{D}_{F_* X} F_* W.$$

Now let $e_1, \ldots, e_n$ be the usual orthonormal fields on R^n and define C^∞ functions b_{rs} on M by $F_*(e_s)_p = \Sigma_{r=1}^n b_{rs}(p)(e_r)_{p'}$. The functions b_{rs} are C^∞ since F, M, and M' are C^∞, and the n by n matrix $b_{rs}(p)$ is orthogonal since F is an isometry. Then for any tangent vector X to M at any point p in M, we know $\overline{D}_X e_s = 0$, since e_s are parallel fields on R^n. Thus,

$$0 = F_*(\overline{D}_X e_s) = \overline{D}_{F_* X}(F_* e_s)$$

$$= \Sigma_{r=1}^n [(Xb_{rs})e_r + b_{rs}\overline{D}_{F_* X} e_r] = \Sigma_r (Xb_{rs})e_r,$$

so $Xb_{rs} = 0$ for all r and s. Since X and p were arbitrary and M is connected, the functions b_{rs} are constant on M and thus the Jacobian of F is a constant orthogonal transformation relative to the natural base $e_1,\ldots,e_n$ of R^n.

Next define a map G of R^n onto itself which is a translation followed by an orthogonal map by letting $G(p) = p' = F(p)$ for one p in M and requiring $(G_*)_p = (F_*)_p$. This completely determines G and the Jacobian of G is constant and hence equal to the Jacobian of F at all points. Since M is connected, $F = G|_M$.//

Section 6.6. Cartan viewpoint and coordinate viewpoint.

In this section let M be a hypersurface of a Riemannian n-manifold $\bar{M}$. Let p be in M, let $\bar{U}$ be a special coordinate neighborhood of p in $\bar{M}$ with U the corresponding neighborhood of p in M, $U \subset \bar{U}$. Apply the Gram-Schmidt process to the coordinate vector fields on $\bar{U}$ to obtain an orthonormal base field $e_1,\ldots,e_n$ on $\bar{U}$ with $e_1(m),\ldots,e_{n-1}(m)$ a base of M_m for m in U and $e_n(m)$ normal to M_m (thus e_n provides a local normal for the neighborhood U). Let f: $U \to \bar{U}$ be the inclusion map.

Applying the results of section 5.2, let $\bar{w}_1,\ldots,\bar{w}_n$ be the dual 1-forms associated with $e_1,\ldots,e_n$ and let $\bar{w}_{ij}$ for $1 \leq i, j \leq n$ be the connexion 1-forms associated with the Riemannian connexion $\bar{D}$ on $\bar{U}$, so

$$\bar{D}_X e_j = \Sigma_{i=1}^{n} \bar{w}_{ij}(X) e_i, \tag{18}$$

for $j = 1,\ldots,n$.

Let $w_{ij} = \bar{w}_{ij}|_U$ and $w_i = \bar{w}_i|_U$ for $1 \leq i, j \leq n$, i.e., $w_{ij} = f^*\bar{w}_{ij}$, $w_i = f^*\bar{w}_i$. Then, if X is tangent to M at m in U, by the Gauss equation,

$$D_X e_j = \Sigma_{i=1}^{n-1} \bar{w}_{ij}(X) e_i \tag{19}$$

$$V(X, e_j) = \bar{w}_{nj}(X) e_n \tag{20}$$

for $j = 1,\ldots,n-1$. Thus w_{ij} for $i, j \leq n$ are the connexion forms for the induced Riemannian connection D on M. Moreover,

(21) $$L(X) = \bar{D}_X e_n = \Sigma_{i=1}^{n-1} w_{in}(X) e_i$$

since $L(X)$ in M_m, so $w_{nn} = 0$ on U. Also $w_n = 0$ on U, since e_n is normal to M. Equation (18) is the Gauss equation and equation (21) is the Weingarten equation.

Let I, II, III be the first, second, and third fundamental forms, respectively. Then for X and Y in M_m, m in U,

$$I(X, Y) = \Sigma_1^{n-1} w_i(X) w_i(Y)$$

$$II(X, Y) = \langle LX, Y \rangle = \Sigma_1^{n-1} w_{in}(X) w_i(Y)$$

$$III(X, Y) = \langle LX, LY \rangle = \Sigma_1^{n-1} w_{in}(X) w_{in}(Y).$$

Notice $0 = X \langle e_i, e_j \rangle = \langle D_X e_i, e_j \rangle + \langle e_i, D_X e_j \rangle = w_{ji}(X) + w_{ij}(X)$ for all X tangent to M, i.e. $w_{ji} = -w_{ij}$ for connexion forms belonging to an orthonormal base (and this again shows $w_{nn} = 0$). Thus we can write II and III in terms of w_{ni} if we wish.

Certain relations are implied by the Cartan structural equations. The equation $d\bar{w}_n = -\Sigma_{j=1}^{n} \bar{w}_{nj} \wedge \bar{w}_j = 0$ (on M_m) implies II is symmetric. The equation $d\bar{w}_{nn} = -\Sigma_{j=1}^{n} \bar{w}_{nj} \wedge \bar{w}_{jn} = 0$ (on M_m) implies III is symmetric. For $i, j \leq n$, $d\bar{w}_{ij} = -\Sigma_{s=1}^{n} \bar{w}_{is} \wedge \bar{w}_{sj} + \bar{R}_{ij}$ when restricted to vectors on M_m, gives $f^* d\bar{w}_{ij} = dw_{ij} = -\Sigma_{s=1}^{n-1} w_{is} \wedge w_{sj} + R_{ij} = -\Sigma_{s=1}^{n} w_{is} \wedge w_{sj} + \bar{R}_{ij}$. Thus

(22) $$R_{ij} = -w_{in} \wedge w_{nj} + \bar{R}_{ij}$$

which is the Gauss curvature equation from this point of view. For $i \leq n$,

(23) $$f^* d\bar{w}_{in} = dw_{in} = -\Sigma_{s=1}^{n-1} w_{is} \wedge w_{sn} + \bar{R}_{in}$$

is the Codazzi-Mainardi equation.

For the coordinate viewpoint, let $x_1, \ldots, x_n$ be the special coordinate system on $\bar{U}$ such that $x_1, \ldots, x_{n-1}$ give coordinates on U. Let $X_i = \partial/\partial x_i$ for $i = 1, \ldots, n-1$ and let $X_n = e_n$ the unit normal (on U). Now apply the above analysis to the base field $X_1, \ldots, X_n$ (and this time

$w_{ij} \neq -w_{ji}$ necessarily, since the base $X_1,\ldots,X_n$ is not necessarily orthonormal).

Section 6.7. Canonical spaces of constant curvature.

We exhibit the three classical examples of n-dimensional ($n \geq 2$) simply connected complete spaces with constant Riemannian curvature $K = 0$, $K > 0$, and $K < 0$; i.e., the Riemannian curvature $K(P)$ of all plane sections is a constant.

For $K = 0$, let $M = R^n$ with the usual Riemannian metric. This is usually called *Euclidean space* or *flat space*.

For $K > 0$, let $M = [a \text{ in } R^{n+1}\colon \Sigma_1^{n+1} a_i^2 = 1/K]$, i.e., M is the n-dim sphere of radius $1/\sqrt{K}$ about the origin in R^{n+1}. It is a Riemannian manifold via the induced metric from R^{n-1}. This is called *spherical space* or *Riemann space*. Letting N be the unit outer normal on M, then $L(X) = \sqrt{K}X$ for all vectors tangent to M and all points are umbilic. By equation (17) above, $K(P) = \langle LX, X\rangle\langle LY, Y\rangle = K$ where X and Y are unit orthogonal vectors spanning P. Since M is compact, it is complete. An alternate proof that M has constant curvature is provided by the group of orthogonal tranformations on R^{n+1}, which provides isometries that will map any point m and plane section P at m, into any other point m' and plane section P'. Since an isometry preserves the curvature, this would show M has constant Riemannian curvature but would not evaluate this constant.

For $K < 0$, let $M = [a \text{ in } R^n\colon \Sigma_1^n a_i^2 < -4/K]$. Let $x_1,\ldots,x_n$ be the usual coordinate functions on R^n, i.e., $x_i(a) = a_i$, let $X_i = \partial/\partial x_i$ for $i = 1,\ldots,n$, and define a metric on M by the functions $g_{ij} = \langle X_i, X_j\rangle = \delta_{ij}/A^2$ where $A = 1 + (K/4)\Sigma_1^n x_r^2$. Then M with this metric is called *hyperbolic space*, or *Poincare space*. Thus M is obtained by a conformal change of the usual metric tensor on an open ball in R^n, and M is simply connected, since it is contractible.

One proves M has constant negative Riemannian curvature K by a direct computation which we outline. Let K_{ij} be the Riemannian curvature of the plane section spanned by X_i, X_j at any point in M. Let $R(X_i, X_j)X_r = \Sigma_k R_r^k(X_i, X_j)X_k = \Sigma_k R_{rij}^k X_k$ define functions R_{rij}^k. Then $K_{ij} = A^2 R_{jij}^i$, and compute via the classical formulae for R_{rij}^k in terms of Γ_{jk}^i, and Γ_{jk}^i in terms of g_{ij}. These formulae show $\Gamma_{jk}^i = 0$ unless two indices are equal and $\Gamma_{ij}^i = \Gamma_{ji}^i = \Gamma_{ii}^i = -Kx_i/2A$ while $\Gamma_{ii}^j = Kx_j/2A$. Then $R_{jij}^i = (K/A) - K^2(\Sigma_1^n x_r^2)/4A^2$ and $K_{ij} = K$.

Also by direct computation one shows $R^i_{jkr} = 0$ unless $k = i$, $r = j$, or $k = j$, $r = i$. Then letting $e_i = AX_i$ gives an orthonormal base e_1, ..., e_n at each point of M. Let P be any plane section at m in M and let f_1, f_2 be an orthonormal base of P which we extend to an orthonormal base of M_m. Then the base e_i is related to the base f_j via an orthogonal matrix, and one uses this fact to show $K(f_1, f_2) = K$. Thus M has constant negative curvature.

To show M is complete, let $K = -B^2$, and one shows the curve g: $t \to [2 \sinh \frac{B}{2}t/(B \cosh \frac{B}{2}t), 0,\ldots,0]$ is a geodesic defined for all t and parameterized by arc length. Such a geodesic is obtained on every ray emanating from the origin 0 by symmetry. Thus

$$\overline{B}_M(0, t) = \overline{B}_{R^n}(0, 2 \sinh \tfrac{B}{2}t/B \cosh \tfrac{B}{2}t)$$

which is a compact set, so M is complete. (Here $\overline{B}_m(p, r) = [m \text{ in } M: d_M(m, p) \leq r$, where d_M is the distance function in $M]$.) Note that the mapping g, when generalized to all rays in R^n, exhibits explicity the exponential map of M_0 onto M (see section 9.3).

For $K > 0$, let $M = R^n$, and let $g_{ij} = \delta_{ij}/A^2$ define a Riemannian metric on M as above. The above computations show M has constant Riemannian curvature K and M is trivially simply connected. But M is not complete since $\overline{B}_M(0, 2\pi/\sqrt{K}) = R^n$ is not compact. Thus we have an example of a conformal change of metric which changes a complete Riemannian manifold into a non-complete Riemannian manifold.

Section 6.8. Existence.

The objective of this section is to show a paracompact connected Hausdorff C^∞ manifold admits a Riemannian metric. This is accomplished by constructing a "partition of the unit function." The function $e^{-(1/x^2)}$ is the principal tool which is used to show there are "many" C^∞ functions on a C^∞ manifold.

LEMMA 1. *If b and c are real numbers with $0 < b < c$. then there exists a C^∞ function f: $R \to R$ with $f(t) = 0$ for $t \leq b$, $0 \leq f(t) \leq 1$ for all t, and $f(t) = 1$ for $t \geq c$.*

Proof. Consider the C^∞ function g which is identically zero for $x \leq 0$ and $e^{-(1/x^2)}$ for $x > 0$. We outline a sequence of operations which leads to the desired functions, and we illustrate (and number) the graphs of these intermediate functions in Fig. 6.1. Translate g so its graph moves $(c - b)/2$ units to the left (this is no. (1)). Reflect the graph of (1) about the y-axis to obtain (2). Multiply (1) and (2) to obtain (3). Integrate (3) to obtain (4). Multiply (4) by a scale factor to obtain (5). Translate the graph of (5) to the right to obtain the desired function f.//

LEMMA 2. *If b and c are real numbers with $0 < b < c$, then there exists a C^∞ function F: $R^n \to R$ with $F(p) = 0$ for $|p| \leq b$, $0 \leq F(p) \leq 1$ for all p, and $F(p) = 1$ for $|p| \geq c$.*

Proof. Let $F(p) = f(|p|)$ where f is obtained from lemma 1. //

LEMMA 3. *If M is a Hausdorff C^∞ manifold and m in M, then there is a coordinate neighborhood U of m and a C^∞ function f: $M \to R$ such that $f(p) > 0$ for p in U and $f(p) = 0$ for p not in U.*

Proof. Let V be any coordinate neighborhood of m with coordinate map ϕ: $V \to R^n$ such that $\phi(m)$ is the origin 0. Choose real numbers b and c with $0 < b < c$ such that $B(0, c) \subset \phi(V)$. Apply lemma 2 to obtain F and let $G = 1 - F$. Then let $U = \phi^{-1}(B(0, c))$ and let $f = G \circ \phi$ on U while $f(p) = 0$ for p not in U.//

LEMMA 4. *If M is a paracompact Hausdorff C^∞ manifold then there exists a locally finite covering $[U_\alpha]$, where U_α are open coordinate neighborhoods, and a collection of non-negative real valued C^∞ functions $[g_\alpha]$ such that $g_\alpha(p) = 0$ for p not in U_α and $\Sigma_\alpha g_\alpha = 1$. The collection $[g_\alpha]$ is called a* partition of unity *for the covering $[U_\alpha]$.*

Proof. Combining lemma 3 and the definition of paracompactness, one obtains the desired covering $[U_\alpha]$ with C^∞ functions f_α: $M \to R$ such that $f_\alpha > 0$ on U_α and $f_\alpha = 0$ on $M - U_\alpha$. The function $F = \Sigma_\alpha f_\alpha$ is a well-defined non-vanishing C^∞ function on M since the covering $[U_\alpha]$ is locally finite. Finally let $g_\alpha = f_\alpha/F$.//

THEOREM. *If M is a connected Hausdorff C^∞ manifold then each of the following three properties implies the other two:*

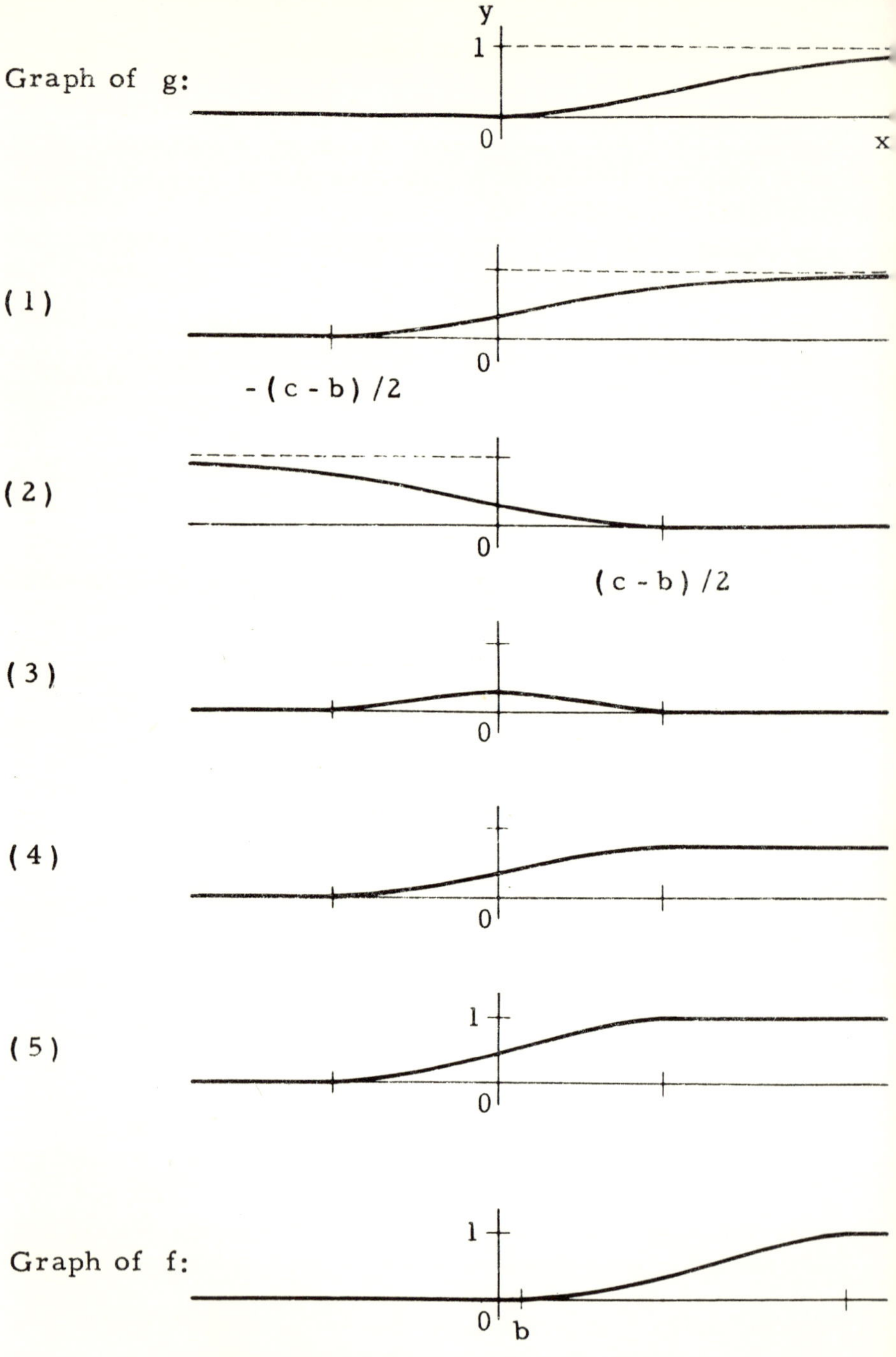

Fig. 6.1 Constructing a C^{∞} Step Function

(a) *M is paracompact,*

(b) *M admits a Riemannian metric,*

(c) *M is second-countable (completely separable).*

Proof. We show (a) implies (b) and give references for the other implications whose proofs are purely topological.

Assuming (a), apply lemma 4 to obtain a locally finite cover $[U_\alpha]$ with the partition of unity $[g_\alpha]$. On each coordinate neighborhood U_α, define a local Riemannian metric tensor $\langle\ ,\ \rangle_\alpha$ by demanding the coordinate map be an isometry. Then the tensor $g_\alpha\langle\ ,\ \rangle_\alpha$ is a global C^∞ tensor on M that vanishes outside U_α. At any point m in M, for X and Y in M_m, let $\langle X, Y\rangle = \Sigma_\alpha g_\alpha(m)\langle X, Y\rangle_\alpha$. This defines a C^∞ Riemannian metric tensor on M which shows (a) implies (b).

Assuming (b), then from section 2.6 we know M is a metric space and hence must be paracompact (see Kelley, p. 160). Thus (b) implies (a). That (c) implies (a) follows from Hocking and Young, p. 79. To show (b) implies (c), we refer the reader to Chapter 6 in Kelley. The metric can be used to define a uniform structure on M which must admit a countable base (see Kelley, p. 186).// For theorems concerning the imbedding of manifolds in other manifolds see Sternberg, Auslander and Mackenzie, or Smale[2].

Problems. All manifolds will be Riemannian unless otherwise stated.

49. If M is semi-Riemannian and D satisfies (4), then D is *metric preserving*. Show that D is metric preserving iff for parallel fields Y and Z along a curve σ for function $\langle Y, Z\rangle$ is constant on σ.

50. Let R and R' be two linear map valued skew-symmetric 2-covariant tensors whose corresponding K and K' satisfy properties (a) thru (d) on p. 124. Show $K = K'$ iff $R = R'$.

51. If f is a C^∞ strictly conformal map, show f_* has no kernel and preserves angles. If f is a complex analytic map, show $\langle f_*X_p, f_*Y_p\rangle = |f'(p)|^2\langle X_p, X_p\rangle$, where f: $C \to C$.

52. Let f: $M \to M'$ be a strictly conformal map with scale function F. Show f is (Riemannian) connexion preserving iff F is

constant and $f(M)$ is flat. If f is an isometry, show f preserves the curvature tensor and the Riemannian curvature.

53. With the standard hypothesis of section 3.3, show if f is connexion preserving, then M is a sphere, a plane, or a right circular cylinder (see Hicks[2]).

54. Let M be a hypersurface in R^n, let N be a C^∞ unit normal, let g be in $C^\infty(M, R)$, and let f_t: $M \to R^n$ by $f_t(p) = p + tg(p)N_p$. Show that $(f_t)_* X = X + t(Xg)N + tgL(X)$ for X tangent to M. If f_t is an isometry for $t > 0$, show that M is flat.

55. Generalize the first two theorems in section 6.5 to the case of a k-submanifold that is framed in an n-manifold for $1 < k < n$ (see Hicks[1]). In the second theorem, if $k = 2$ and $n = 3$, show $\overline{K}(P) = K(P)$ iff g is a line of curvature on M.

56. If u and v are orthonormal coordinates with domain A on a 2-manifold (thus $\partial/\partial u$ and $\partial/\partial v$ are orthonormal), show the coordinate curves are geodesics and $K \equiv 0$ on A.

57. (K. Leisenring) Show that $f(u, v) = (\cos u \cos v, \cos u \sin v, \sin u \cos v, \sin u \sin v)$ is an isometric imbedding of the flat torus T into the unit sphere S^3 in R^4. Show the total (imbedded) curvature of $f(T)$ in S^3 is a constant negative one.

58. Let M be connected with symmetric connexion D and let L_p: $M_p \to M_p$ be a C^∞ linear map valued function on M. If $\mathrm{Tor}_L \equiv 0$ and all points are L-umbilic, show L is a constant multiple of the identity map.

59. Show that every isometry of R^n can be factored uniquely into an orthogonal map followed by a translation. If f: $R^n \to R^n$ is orthogonal, show $f_* = f$ in a natural way.

60. If $e_1, \ldots, e_n$ is an orthonormal base field with dual base $w_1, \ldots, w_n$ and M has constant Riemannian curvature K show the associated curvature forms $R_{ij} = Kw_i \wedge w_j$.

61. If M has constant Riemannian curvature K and one defines a metric on $M \times M$ by $\langle (X_1; Y_1), (X_2; Y_2) \rangle = \langle X_1, X_2 \rangle + \langle Y_1, Y_2 \rangle$, does $M \times M$ have constant curvature?

62. If $x_1,\ldots,x_n$ are coordinates on a hypersurface U in R^{n+1}, let $X_i = \partial/\partial x_i$, $g_{ij} = \langle X_i, X_j\rangle$, $b_{ij} = \langle LX_i, X_j\rangle$, and $LX_j = \Sigma_i a_{ij} X_i$. Show that $a_{ij} = \Sigma_r (g^{-1})_{ir} b_{rj}$ (Weingarten equation), $R^i_{jkh} = \Sigma_r (g^{-1})_{ir}(b_{hj}b_{rk} - b_{kj}b_{rh})$ (Gauss-curvature equation), and

$$\frac{\partial b_{ir}}{\partial x_s} - \frac{\partial b_{is}}{\partial x_r} = \Sigma_r (b_{kr}\Gamma^k_{is} - b_{ks}\Gamma^k_{ir})$$

(Codazzi-Maniardi equation).

63. If M is a Hausdorff C^∞ manifold, A is a compact subset of M, B is open in M, and $A \subset B$, show there is an f in $C^\infty(M, R)$ with $f(A) = 0$, $f(M - B) = 1$, and $0 \leq f(M) \leq 1$.

7. Operators on Forms and Integration

This chapter develops more structure on a manifold. To conserve space, the treatment is fairly blunt and many computational details are omitted. In the first four sections M is a C^∞ n-manifold and A is an open set in M.

Section 7.1. Exterior derivative.

For $p \geq 0$ we define the *exterior differentiation* map d: $F^p(A) \to F^{p+1}(A)$ where $F^p(A)$ is the set of C^∞ p-forms on A. If f in $F^0(A)$ and X is a C^∞ field on A, then $df(X) = Xf$. For $p > 1$, letting w be a $(p-1)$ form on A and $X_1,\ldots,X_p$ be C^∞ fields on A, then

$$dw(X_1,\ldots,X_p) = \Sigma_{j=1}^p (-1)^{j+1} X_j W(X_1,\ldots,\hat{X}_j,\ldots,X_p) + \tag{1}$$

$$\Sigma_{i<j}(-1)^{i+j} w([X_i, X_j], X_1,\ldots,\hat{X}_i,\ldots,\hat{\hat{X}}_j,\ldots,X_p),$$

where $\hat{X}$ indicates that the field X is omitted as an argument.

Notice that the definition is consistent with the partial definition in section 5.2. One proves that dw is in $F^p(A)$ by using the characterization theorem in Chapter 4. We outline the argument. That dw is linear with respect to addition is trivial. That dw is alternating can be shown by switching two arguments and examining the terms that don't immediately change signs (this must be done carefully). That dw is linear over the ring $F^0(A)$ then need only be checked in one slot.

Proposition. The operator d has the following properties:

(1) $d(w + v) = dw + dv$, where w and v are *in* $F^p(A)$.

(2) $d(w \wedge v) = ((dw) \wedge v) + (-1)^p(w \wedge dv)$, for w in $F^p(A)$ and v any form on A. (Any operator with this property is called an *anti-derivation.*)

(3) $d^2 = d \circ d = 0$.

Proof. Property (1) follows trivially from the definitions of d and addition of functionals. For the other two properties we first obtain a local representation of d. Let $x_1,\ldots,x_n$ be a coordinate system on an open set U, and let $X_i = \partial/\partial x_i$. Then on U, a $(p - 1)$-form w may by represented by $w = \Sigma a_{i_1\ldots i_{p-1}} dx_{i_1} \wedge\ldots\wedge dx_{i_{p-1}}$, where the sum is over all indices such that $1 \leq i_j \leq n$ and $i_1 < i_2 <\ldots< i_{p-1}$, and $a_{i_1\ldots i_{p-1}} = w(X_{i_1},\ldots,X_{i_{p-1}})$. $dw = \Sigma da_{i_1\ldots i_{p-1}} \wedge dx_{i_1} \wedge\ldots\wedge dw_{i_{p-1}}$, which is proved by applying both sides to $(X_{k_1},\ldots,X_{k_p})$ for $k_1 < k_2\ldots< k_p$. Since $[X_r, X_s] = 0$, $dw(X_{k_1},\ldots,X_{k_p}) = \Sigma_{j=1}^{p}(-1)^{j+1}X_{k_j}a_{k_1\ldots\hat{k}_j\ldots k_p}$, while $[\Sigma da_{i_1\ldots i_{p-1}} \wedge dx_{i_1} \wedge\ldots\wedge dx_{i_{p-1}}](X_{k_1},\ldots,X_{k_p}) = da_{k_2\ldots k_p}(X_{k_1}) - da_{k_1\hat{k}_2\ldots k_p}(X_{k_2}) +\ldots = dw(X_{k_1},\ldots,X_{k_p})$.

To prove property (2), first let f and g be functions in $F^0(A)$ and note $d(fg) = (df)g + f(dg)$ follows from the derivation property of vectors. Next observe that because of (1) and the local representation above, one need only verify (2) for forms of the type $w = fdx_1 \wedge\ldots\wedge dx_p$ and $v = gdy_1 \wedge\ldots\wedge dy_r$, where x_i and y_i are functions chosen from the members of a coordinate system. Then $w \wedge v = fgdx_1 \wedge\ldots\wedge dx_p \wedge dy_1 \wedge\ldots\wedge dy_r$, and $d(w \wedge v) = d(fg) \wedge dx_1 \wedge\ldots\wedge dy_r = (gdf + fdg) \wedge dx_1 \wedge\ldots\wedge dy_r = dw \wedge v + (-1)^p w \wedge dv$.

For property (3) we first show $d^2 f = 0$ for a C^∞ function f in $F^0(A)$. Locally, $df = \Sigma_{j=1}^{n}(\partial f/\partial x_j)dx_j$ so $d^2 f = \Sigma_{i,j=1}^{n}(\partial^2 f/\partial x_i \partial x_j)dx_i \wedge dx_j = \Sigma_{i<j}[(\partial^2 f/\partial x_i \partial x_j) - (\partial^2 f/\partial x_j \partial x_i)]dx_i \wedge dx_j = 0$. For any w we may represent dw locally as a sum of products of df's for functions f; hence by (2) each term in $d^2 w$ has a factor $d^2 f = 0$, so $d^2 w = 0$.//

Letting $F(M) = \Sigma_{k=0}^{n} F^k(M)$ be the direct sum of the modules of forms of homogeneous type, endowed with its exterior multiplication structure and exterior derivative operator d, one obtains a graded differential algebra which is called the *Cartan differential algebra* of M. If $f\colon M \to N$ is C^∞, then $d \circ f^* = f^* \circ d$ on $F(N)$, and it is sufficient to check this only on 0-forms and 1-forms.

There are other ways to define d, indeed one natural way is to define d via a local representation, get the desired properties, and then show it is independent of the local representation (see Chevalley[1], p. 146). Then the invariant formula we took as definition must be verified. Our treatment in this and the following sections is similar to that of Palais[1].

Section 7.2. Contraction.

Let X be a C^∞ vector field on the open set A. An operator C_X, called *contraction by X*, which maps $F^p(A)$ into $F^{p-1}(A)$ is defined as follows: (a) if f in $F^0(A)$, let $C_X f = 0$, and (b) if w in $F^p(A)$ for $p > 0$, let $(C_X w)(X_1, \ldots, X_{p-1}) = w(X, X_1, \ldots, X_{p-1})$.

Proposition. The operator C has the following properties:

(1) $(C_X)^2 = 0$.

(2) $C_X(w + v) = C_X w + C_X v$,

(3) $C_{X+Y} = C_X + C_Y$,

(4) $C_{fX} = fC_X$,

(5) $C_X(w \wedge z) = (C_X w) \wedge z + (-1)^p(w \wedge C_X z)$,

for f in $F^0(A)$, X and Y in $T^{1,0}(A)$, w and v in $F^p(A)$, and z in $F^q(A)$.

Proof. Properties (1) through (4) are trivial. Property (5) follows by induction on p, and it is sufficient to prove it when w is a product of p 1-forms by the local representation of forms.//

The operator C_X can be defined on covariant tensors and mixed tensors in an obvious way (with only (2), (3), and (4) valid in general), and one can let C_X be zero on pure contravariant tensors. Properties (3) and (4) indicate C is a tensor map (an anti-derivation valued 1-form of degree -1 on $F(A)$).

There is another form of "contraction" induced by the natural identification of tensors of type 1, 1 and linear maps. Let W be an n-dim vector space over R. For $r > 0$, $s > 0$, $1 \le i \le r$, $1 \le j \le s$ define $tr^{i,j}$: $T^{r,s}(W) \to T^{r-1,s-1}(W)$ by taking θ in $T^{r,s}(W)$, $w_1,\dots,w_{r-1}$ in W^*, and $X_1,\dots,X_{s-1}$ in W and letting

$$(2) \qquad (tr^{i,j}\theta)(w_1,\dots,w_{r-1}, X_1,\dots,X_{s-1}) =$$

$$\Sigma_{k=1}^{n}\theta(w_1,\dots,w_{i-1}, z_k, w_i,\dots,w_{r-1}, X_1,\dots,X_{j-1}, Z_k, X_j,\dots,X_{s-1}$$

where $Z_1,\dots,Z_n$ is a base of W and $z_1,\dots,z_n$ the dual base of W^*. One checks easily that $tr^{i,j}\theta$ is well-defined independently of the particular base used. If θ in $T^{1,1}(W)$, let $tr^{1,1}\theta = tr\theta$. The above operator induces an operator $tr^{i,j}$: $T^{r,s}(A) \to T^{r-1,s-1}(A)$ for an open set A in M.

Section 7.3. Lie derivative.

Let X be a C^∞ vector field on the open set A. An operator L_X, called the *Lie derivative via X*, which maps $T^{r,s}(A)$ into itself, is defined as follows: (a) if f in $F^0(A)$, $L_Xf = Xf$; (b) if Y in $T^{1,0}(A)$, $L_XY = [X, Y]$; (c) if w in $T^{0,1}(A)$, $(L_Xw)(Y) = Xw(Y) - w([X, Y])$; and (d) if θ in $T^{r,s}(A)$, $w_1,\dots,w_r$ in $T^{0,1}(A)$, and $Y_1,\dots,Y_s$ in $T^{1,0}(A)$, then $L_X\theta$ is defined by solving for it in the equation

$$(3) \qquad L_X[\theta(w_1,\dots,w_r, Y_1,\dots,Y_s)] = (L_X\theta)(w_1,\dots,Y_s)$$

$$+ \theta(L_Xw_1,w_2,\dots,Y_s) +\dots+\theta(w_1,\dots,Y_{s-1}, L_XY_s).$$

We call L_X a *complete derivation* because of the property (d), and note all terms in (3) are well-defined by (a), (b), and (c) except the $L_X\theta$ term (indeed, (c) is "defined" by (d)). One shows $L_X\theta$ is a tensor by checking the linearity over $F^0(A)$.

Proposition. The operator L_X has the following properties:

(1) L_X preserves forms,

(2) $L_X(w + z) = L_Xw + L_Xz$,

(3) $L_X(w \otimes v) = (L_X w) \otimes v + w \otimes L_X v,$

(4) $L_X(\alpha \wedge \beta) = (L_X \alpha) \wedge \beta + \alpha \wedge L_X \beta,$

where w and z are tensors of the same type, v is any tensor, and α and β are forms.

Proof. An exercise (for (4) use $L_X[(\alpha \otimes \beta)^\pi] = [L_X(\alpha \otimes \beta)]^\pi$).

There is a more geometric definition of the Lie derivative L_X on covariant tensors which we now discuss. Suppose the vector field X is defined and C^∞ on all of M. For each m in M let $f_m(t)$ be the integral curve of X (section 1.5) through m with $f_m(0) = m$. we know f_m defined for t in a neighborhood of zero, but *suppose* each f_m is defined for all t and R. Then for each t in R we could define a map F_t: $M \to M$ by $F_t(m) = f_m(t)$, with the properties $F_t \circ F_s = F_{t+s}$ and F: $M \times R \to M$ by $F(m, t) = F_t(m)$ would be C^∞ (from the fact that X was C^∞ and the C^∞ dependence of solutions of ordinary differential equations upon initial conditions). Each F_t would be a diffeo, since $(F_t)^{-1} = F_{-t}$ and F_0 is the identity map. A map F with the above properties is called a 1-*parameter group of differentiable transformations of M*, and X is called its *infinitesimal generator*.

In general f_m is not defined for all t, but one does obtain a local 1-*parameter group of local transformations* in a neighborhood of each m in M; i.e., for each m in M there is a neighborhood U of m, a real number $b > 0$, and a map F: $U \times (-b, b) \to M$ such that (1) F is C^∞, (2) for t in $(-b, b)$, F_t: $U \to F_t(U)$ is a diffeo, (3) for t, s, and $t + s$ in $(-b, b)$, $F_t \circ F_s = F_{t+s}$, and (4) for fixed p in U, $f_p(t) = F_t(p)$ is an integral curve of X. For more details see Palais[1 and 2] and Nomizu[1] (p. 5).

LEMMA 1· *Let Y be a C^∞ field in a neighborhood of m in M. We choose U and b in the preceding paragraph to be sufficiently small so the image of F is contained in the domain of Y. Then*

$$[X, Y]_m = \lim_{t \to 0} [(F_{-t})_* Y_{F(m,t)} - Y_m]/t.$$

Proof. See Nomizu[1] (p. 8).

Assuming lemma 1, which gives us another geometric interpretation of the bracket, it is trivial to show the following lemma .

LEMMA 2. *Let w be a C^∞ p-form at m. Then*

$$(L_X w)_m = \lim_{t\to 0} [(F_t^* w)_m - w_m]/t$$

where

$$(F_t^* w)_m(Y_1,\dots,Y_p) = w_{F(m,t)}((F_t)_* Y_1,\dots,(F_t)_* Y_p).$$

The following is a useful relation between d, L_X, and C_X.

Proposition. If X is a C^∞ field on A, then $L_X = d \circ C_X + C_X \circ d$ when applied to C^∞ forms on A.

Proof. We verify this equality on function (0-forms) and 1-forms. This is sufficient to prove the proposition, since locally a form is a sum of products of functions and 1-forms, and the operators which we equate above are both derivations; hence their value on any form is determined by the values on functions and 1-forms.

For f in $F^0(A)$, $dC_X(f) + C_X d(f) = 0 + df(X) = Xf = L_X f$. For w in $F^1(A)$, $(dC_X w + C_X dw)(Y) = Yw(X) + dw(X, Y) = Yw(X) + Xw(Y) - Yw(X) - w([X, Y]) = (L_X w)(Y)$.//

Section 7.4. *General covariant derivative.*

Let D be a connexion on M, and let X be a C^∞ field on the open set A. An operator D_X, called the *covariant derivative via X*, which maps $T^{r,s}(A)$ into itself, is defined by using the recipe for defining L_X. The definition of D_X proceeds exactly as the definition for L_X except for (b), and if Y in $T^{1,0}(A)$, $D_X Y$ is given by the connexion D (see section 5.1).

When D_X is substituted for L_X in the first proposition of the previous section, one obtains valid properties for D_X.

An operator Δ, called the *general covariant derivative operator*, which maps $T^{r,s}(A)$ into $T^{r,s+1}(A)$ is induced by D. If θ is in $T^{r,s}(A)$, $w_1,\dots,w_r$ are in $T^{0,1}(A)$, and $Y_1,\dots,Y_{s+1}$ are in $T^{1,0}(A)$, then

(4) $(\Delta\theta)(w_1,\ldots,w_r, Y_1,\ldots,Y_{s-1}) = (D_{Y_{s+1}}\theta)(w_1,\ldots,w_r, Y_1,\ldots,Y_s).$

That $(\Delta\theta)$ is a tensor is left as a problem. If θ and ϕ are tensors of the same type, then $\Delta(\theta + \phi) = \Delta\theta + \Delta\phi$, but Δ is not a tensor (see problem 64).

If $1 \leq i \leq p$ and $1 \leq j \leq q$, then

(5) $\Delta \circ tr^{i,j} = tr^{i,j} \circ \Delta$

on $T^{p,q}(A)$.

An operator div, called the *divergence*, which maps $T^{r,s}(A)$ into $T^{r-1,s}(A)$, for $r > 0$ and $s \geq 0$, is defined by div $= tr^{r,s+1} \circ \Delta$. We write div $\theta = tr(\Delta\theta)$, where we assume the trace is taken on the last covariant slot and the last contravariant slot. A tensor θ is *conservative* if div $\theta = 0$.

The *Riemann-Christoffel curvature tensor* of type 1, 3 is the tensor K in $T^{1,3}(A)$ defined by

(6) $K(w, X, Y, Z) = w(R(Y, Z)X)$

for w in $T^{0,1}(A)$ and X, Y, and Z in $T^{1,0}(A)$. The *second Bianchi identity* is the equation

(7) $(\Delta K)(w, X, Y, Z, W) + (\Delta K)(w, X, W, Y, Z,) + (\Delta K)(w, X, Z, W, Y,) = 0$

which is valid if D is symmetric, and it is proved by noting the expression

(8) $D_W(R(Y, Z)X) - R(Z, [Y, W])X - R(Y, Z)(D_W X),$

when written on three lines, permuting W, Y, Z cyclically from line to line, and then adding the three lines, yields zero.

The *Ricci tensor* is the 2-covariant tensor

(9) $\text{Ric}(X, Y) = (tr^{1,2}K)(X, Y) = -(tr^{1,3}K)(X, Y)$

(and this is the negative of the "classical" Ricci tensor). Notice $(tr^{1,1}K)(X, Y) = trR(X, Y)$, The *Ricci curvature* of a vector X is the number $\mathrm{Ric}(X, X)$ (and this agrees with the "classical" Ricci curvature). If D is symmetric, the first Bianchi identity implies

(10) $\quad \mathrm{Ric}\,(X, Y) = \mathrm{Ric}\,(Y, X) + trR(X, Y).$

If D is Riemannian, then $R(X, Y)$ is skew-symmetric by (c) in section 6.2, so Ric is symmetric. Hence there exists a self-adjoint linear map R^*, called the *Ricci map*, defined on each M_m with $\mathrm{Ric}\,(X, Y) = \langle R^*(X), Y\rangle$; indeed

(11) $\quad R^*(X) = \Sigma_{j=1}^{n} R(X, Z_j)Z_j$

for an orthonormal base $Z_1,\dots,Z_n$. By (11), R^* is C^∞. The *scalar curvature* $S(m)$ at each m in M is defined by $S(m) = tr(R^*)_m$.

A (semi-) Riemannian metric induces many operations called "raising" and "lowering" of indices which we now explain. The non-singular metric tensor induces a non-singular linear map G of M_m onto M_m^* for each m, i.e., if X in M_m, then $G(X)(Y) = \langle X, Y\rangle$. We let G_* denote the inverse map of M_m^* onto M_m. If w in M_m^*, then $\langle G_*w, X\rangle = w(X)$. If $1 \leq i \leq r$, $1 \leq j \leq s+1$, and θ is in $T^{r,s}$ define $G^{i,j}\theta$ in $T^{r-1,s+1}$ by

(12) $\quad (G^{i,j}\theta)(w_1,\dots,w_{r-1}, X_1,\dots,X_{s+1})$

$$= \theta(w_1,\dots,w_{i-1}, G(X_j), w_i,\dots,w_{r-1}, X_1,\dots,\hat{X}_j,\dots,X_{s+1}).$$

Similarly, define $G_*^{i,j}$: $T^{r,s} \to T^{r+1,s-1}$ for $1 \leq i \leq r+1$ and $1 \leq j \leq s$ by taking the form in the ith covariant slot (of the new tensor); applying G_*, and inserting it into the jth contravariant slot (of the old tensor). Thus $G^{1,1} = G$ on $T^{1,0}$, and the 1,1-tensor $\bar{R}$ associated with R^* is given by $\bar{R} = G_*{}^{1,1}\,\mathrm{Ric}$ (where $\bar{R}(w, X) = w(R^*X)$). If f is in $C^\infty(M, R)$, the *gradient field* of f is the field grad $f = G_*(df)$ and the *Laplacian* of f is the function del f = div(grad f); (sometimes the notation del f = Δf is used).

The operators $G^{i,j}$ and $G_*^{i,j}$ commute with Δ when possible, i.e.,

(13) $\quad \Delta \circ G^{i,j} = G^{i,j} \circ \Delta$ on $T^{r,s}$ if $j \leq s+1$ and

(14) $\Delta \circ G_*^{i,j} = G_*^{i,j} \circ \Delta$ on $T^{r,s}$ if $i \leq r + 1$.

As an example of the use of these operations we prove that

(15) $\Delta S = 2 \operatorname{div} \bar{R}$,

which is used in general relativity. Let $Z_1,\ldots,Z_n$ be an orthonormal base of M_m and $w_1,\ldots,w_n$ the dual base. The second Bianchi identity implies

$$\Sigma_{i,j}[\Delta K(w_j, Z_i, Z_j, Z_i, X) + \Delta K(w_i, Z_i, X, Z_j, Z_i) + K(w_j, Z_i, Z_i, X, Z_j)] =$$

The first term of the sum gives $(\Delta S)(X)$, while the other two each give $-(\operatorname{div} \bar{R})(X)$. For

$$(\Delta S)(X) = (\Delta tr^{1,1}G_*{}^{1,1}tr^{1,2}K)(X)$$

$$= (tr^{1,1}G_*{}^{1,1}tr^{1,2}\Delta K)(X)$$

$$= \Sigma_{ij}\Delta K(w_j, Z_i, Z_j, Z_i, X),$$

$$(\operatorname{div} \bar{R})(X) = (tr^{1,2}\Delta G_*{}^{1,1}tr^{1,2}K)(X)$$

$$= -\Sigma_{i,j}\Delta K(w_j, Z_i, X, Z_j, Z_i), \text{ and}$$

$$(\Delta K)(w_j, Z_i, Z_i, X, Z_j) = (\Delta G^{1,1}K)(Z_j, Z_i, Z_i, X, Z_j)$$

$$= (\Delta G^{1,1}K)(Z_i, X, Z_j, Z_i, Z_j)$$

$$= -(\Delta G^{1,1}K)(Z_i, Z_i, X, Z_j, Z_j) - (\Delta G^{1,1}K)(Z_i, Z_j, Z_i, X, Z_j)$$

$$= -(\operatorname{div} \bar{R})(X),$$

by (c) and (a') in section 6.2.

Section 7.5. Integration of forms and Stokes' theorem.

One integrates p-forms over p-chains, or singular p-chains, which we now define. Let $I^p = [a \text{ in } R^p\colon\ 0 \leq a_i \leq 1]$ denote the *unit p-square* for $p > 0$, and $I^\circ = [0 \text{ in } R]$. A C^∞ *p-cube* σ on M is an M-valued C^∞ function defined on an open neighborhood of the unit p-square I^p in R^p. A *real C^∞ p-chain* c is a finite formal linear combination of C^∞ p-cubes with real coefficients, thus $c = r_1\sigma_1 + r_2\sigma_2 + \dots + r_k\sigma_k$ where r_j in R and σ_j are C^∞ p-cubes. The set $C_p(M, R)$ of all real C^∞ p-chains is an abelian group (actually an R-module) where one defines addition by adding the coefficients of corresponding p-cubes.

There are fancier ways of defining $C_p(M, R)$. Let Q_p be the set of C^∞ p-cubes on M. Then $C_p(M, R)$ is isomorphic to set of all functions mapping Q_p into R which are zero except on a finite number of elements, and the addition and scalar multiplication structure on this function space is obvious. Similarly, one could define $C_p(M, Z)$, the set of *integral C^∞ p-chains* or *C^∞ p-chains over the integers*. Then $C_p(M, R) = R \quad {}_Z\, C_p(M. Z)$. More generally one could define C^∞ p-chains over any ring A with an identity element, and then by using the tensor product obtain the A-module of C^∞ p-chains on M over any A-module. There are corresponding groups obtained from C^r p-chains for any integer $r \geq 0$. These groups are fundamental objects of the *cubical singular homology theory* for M and are studied in algebraic topology, (see Eilenberg and Steenrod). Because of our differential geometry bias, we restrict ourselves to real C^∞ p-chains, and let $C_p = C_p(M, R)$.

The *support of a p-cube* σ is the set $|\sigma| = \sigma(I^p)$, the image of I^p under σ. The *support of a p-chain* c is the set $|c| = U_i|\sigma_i|$ for σ_i in c, where we say σ_i in c if the coefficient of σ_i is non-zero, i.e., adopting the functional viewpoint $c(\sigma_i) \neq 0$ iff σ_i in c.

To define the *boundary map* $\partial\colon C_p \to C_{p-1}$, define maps a_i^1 and a_i^0 from I^{p-1} into I^p for $i = 1,\dots,p$ by

$$(16) \qquad a_i^\epsilon(t_1,\dots,t_{p-1}) = (t_1,\dots,t_{i-1}, \epsilon, t_i,\dots,t_{p-1})$$

where $\epsilon = 1$ or 0. If σ in Q_p, define $\partial\sigma = \Sigma_1^p(-1)^{i+1}(\sigma \circ a_i^1 - \sigma \circ a_i^0)$, and call the $(n-1)$-cubes $\sigma \circ a_i^1$ and $\sigma \circ a_i^0$ *faces of* σ. We extend ∂ to all of C_p by demanding it be linear, i.e., $\partial(c_1 + c_2) = \partial c_1 + \partial c_2$

and $\partial(rc) = r\partial c$ for r in R. A straightforward computation shows $\partial^2 = 0$.

For $p > 0$, let σ be a C^∞ p-cube, let w be a p-form, and let $u_1, \ldots, u_p$ be the natural coordinate function on R^p. Since $\sigma^* w$ is a p-form on a neighborhood of I^p, we may define a C^∞ function f on I^p by $\sigma^* w = f du_1 \wedge du_2 \wedge \ldots \wedge du_p$. Then

$$(17) \qquad \int_\sigma w \equiv \int_{I^p} \sigma^* w \equiv \int_{I^p} f$$

where the integral on the right is the standard Riemann integral of f over I^p developed in advanced calculus. If $c = \Sigma_1^k r_i \sigma_i$ is a p-chain, then $\int_c w = \Sigma_1^k r_i \int_{\sigma_i} w$; thus for fixed w, the integral over w is an R-homomorphism of C_p into R. Since σ^* is linear, it is trivial that $\int_c (w_1 + w_2) = \int_c w_1 + \int_c w_2$ for p-forms w_i and a p-chain c.

For $p = 0$, let f be a function on M and σ_m the 0-cube with $\sigma_m(0) = m$, then $\int_{\sigma_m} f = f(m) = \sigma_m^* f(0)$, and we extend the integral of f over any real 0-chain to be linear (as extended above).

Let $C^p = \mathrm{Hom}_R(C_p, R)$, which is the R-module of all R-linear homomorphisms of C_p into R. The set C^p is called the module of *real C^∞ p-cochains of M*. The adjoint δ of the boundary operator ∂ is called the *coboundary operator* and is defined by $\delta f(c) = f(\partial c)$ for a p-cochain f and a $(p + 1)$-chain c. Thus $\delta\colon C^p \to C^{p+1}$ and $\delta^2 = 0$.

We define the *Stokes' map* $S\colon F^p(M) \to C^p$ which maps p-forms on M into C^∞ p-cochains on M by $[S(w)](c) = \int_c w$, for c in C_p. The following theorem shows the Stokes' map commutes with the differential coboundary operator, i.e., $S \circ d = \delta \circ S$.

STOKES' THEOREM. *Let w be a C^∞ p-form and σ be a $C^\infty(p + 1)$-cube, then*

$$(18) \qquad \int_\sigma dw = \int_{\partial\sigma} w.$$

Proof. Define C^∞ functions $a_1, \ldots, a_{p+1}$ on I^{p+1} by $\sigma^* w = \Sigma_1^{p+1} a_i du_1 \wedge du_2 \wedge \ldots \wedge \widehat{du_i} \wedge \ldots \wedge du_{p+1}$. Then

$$d(\sigma^* w) = \Sigma_{i=1}^{p+1}(\Sigma_{j=1}^{p+1} \frac{\partial a_i}{\partial u_j} du_j) \wedge du_1 \wedge \ldots \wedge \widehat{du_i} \wedge \ldots \wedge du_{p+1}$$

$$= [\Sigma_{i=1}^{p+1}(-1)^{i+1} \frac{\partial a_i}{\partial u_i}] du_1 \wedge \ldots \wedge du_{p+1}.$$

Hence $\int_\sigma dw = \int_{I^{p+1}} \sigma^* dw = \int_{I^{p+1}} d\sigma^* w = \int_{I^{p+1}} [\Sigma_1^{p+1}(-1)^{i+1} \frac{\partial a_i}{\partial u_i}]$

$$= \Sigma_1^{p+1}(-1)^{i+1}[\int_0^1 \int_0^1 \ldots \int_0^1 \frac{\partial a_i}{\partial u_i} du_1 du_2 \ldots du_{p+1}]$$

$$= \Sigma_1^{p+1}(-1)^{i+1} \int_{I^p} (a_i \circ a_i^1 - a_i \circ a_i^0),$$

where we use Fubini's theorem and integrate first with respect to i^{th} coordinate to obtain the last equality.

For the other side we must compute $\int_{\sigma \circ \alpha_i^\varepsilon} w = \int_{I^p} (a_i^\varepsilon)^* \circ \sigma^*(w)$ for $\epsilon = 0$ or 1. Notice $(a_i^\varepsilon)^* du_j = d((a_i^\varepsilon)^* u_j) = d(u_j \circ a_i^\varepsilon) = du_j$, 0, or du_{j-1} according as $j < i$, $j = i$, or $j > i$, respectively. Thus $(a_i^\varepsilon)^* \sigma^* w = (a_i \circ a_i^\varepsilon) du_1 \wedge \ldots \wedge du_p$ and

$$\int_{\partial\sigma} w = \Sigma_{i=1}^{p+1}(-1)^{i+1}[\int_{\sigma \circ \alpha_i^1} w - \int_{\sigma \circ \alpha_i^0} w]$$

$$= \Sigma_{i=1}^{p+1}(-1)^{i+1} \int_{I^p} (a_i \circ a_i^1 - a_i \circ a_i^0),$$

which proves the desired equality.//

We remark that Stokes' theorem is simply a generalized "fundamental theorem of calculus." Let f: $M \to M'$ be a C^∞ map, let w be a p-form on M' and σ a p-cube in M, then it is trivial to show $\int_{f \circ \sigma} w = \int_\sigma f^* w$, which is essentially the classical substitution rule that deals with the behavior of integrals with respect to mappings.

The Stokes' map induces a map at the cohomology level that yields an algebra-isomorphism of the differential cohomology groups of a manifold with the real singular cohomology groups. This fact is called the de Rham theorem (see A. Weil, and problem 71).

Section 7.6. Integration in a Riemannian manifold.

Let M be a Riemannian manifold, let σ be a C^∞ curve in M, and let f be a real valued C^∞ function on the image of σ, i.e., let $f \circ \sigma$ be C^∞. Consider a "piece" of σ, which we assume to be parameterized by arc length on the interval $[a, b]$, and define

$$\text{(19)} \qquad \int_{\sigma|[a,b]} f = \int_a^b f \circ \sigma(s)ds,$$

where $\sigma|[a, b]$ denotes the restriction of σ to the interval $[a, b]$. Call the integral just defined *the integral of f over σ restricted to* $[a, b]$, and when the interval is understood, we write simply $\int_\sigma f$. If f is a C^∞ real valued function f defined on a broken C^∞ curve σ, we define $\int_\sigma f$ to be the sum of the integrals of f over the finite number of C^∞ sub-curves determining σ. Notice that by assuming σ parametrized by arc length we are integrating over *oriented* or *directed* curves.

We wish to integrate real valued C^∞ functions over other subsets of M, and in some cases over M itself. This could be accomplished by using the Riemannian metric to define a measure on M, but for our purposes we need not be so general. First we define orientable manifolds and then utilize the theory developed above for integrating forms over chains.

An n-dimensional manifold M is *orientable* if there is a non-vanishing C^∞ n-form w on M. When M is orientable and we have selected w, we say M is *oriented* (by w) and w is an *orientation* of M. If M is oriented by w, then an ordered base $e_1,\ldots,e_n$ of M_m is *positively oriented* if $w_m = bw_1 \wedge \ldots \wedge w_n$ where $b > 0$ and w_i are the 1-forms dual to e_j. We say M is *non-orientable* when M is not orientable. If M is oriented and $e_1,\ldots,e_n$ a positively oriented base of M_m, then one verifies easily that a base $f_1,\ldots,f_n$ of M_m is positively oriented if and only if $\det(b_{ij}) > 0$ where $f_j = \Sigma_i b_{ij} e_i$.

For example, R^n is orientable, and we orient it by choosing $w = du_1 \wedge \ldots \wedge du_n$ where u_i are the natural coordinate functions. It is a topological result that any complete (or closed) hypersurface in R^n is orientable.

Let M and M' be oriented n-manifolds. A non-singular C^∞ map f of M into M' is *orientation preserving* if f_* maps a positively oriented base onto a positively oriented base.

Let M be an oriented Riemannian n-manifold. For m in M let $e_1, \ldots, e_n$ be a positively oriented orthonormal base of M_m with dual base $w_1, \ldots, w_n$. Define the n-form v by $v_m = w_1 \wedge \ldots \wedge w_n$. The form v is a well-defined (independent of the particular base) C^∞ n-form on M called the *volume element*.

A major problem now confronts us: the problem of "triangulating" or "cubulating" a manifold. This is a theory for breaking the manifold into "nice pieces" over which one can integrate functions. For this purpose we define fundamental n-chains. Let Int(A) denote the interior of a set A.

Let M be an oriented C^∞ n-manifold. A *fundamental n-chain* in M is a chain $c = \sigma_1 + \ldots + \sigma_k$ such that: (1) each σ_i is an n-cube that is an orientation preserving diffeo onto its image; and (2) Int $(|\sigma_i|) \cap$ Int $(|\sigma_j|)$ is empty for $i \neq j$. Figure 7.1 gives a schematic diagram of a fundamental 2-chain (with the images of the faces of the canonical 2-cube numbered).

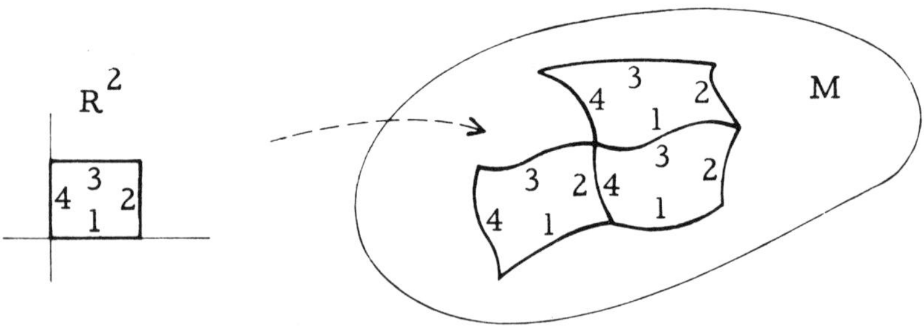

Fig. 7.1. A Fundamental 2-Chain

If M is an oriented Riemannian n-manifold, c is an n-chain, and f is a C^∞ real valued function whose domain contains $|c|$, then define

$$\int_c f = \int_c fv \tag{20}$$

where v is the volume element on M. Let a subset A of M be *fundamental* if there exists fundamental n-chain c with $|c| = A$. Notice a fundamental set is compact.

Proposition. If c and τ are two fundamental n-chains with $|c| = |\tau| = A$, and f is a C^∞ function whose domain contains A, then $\int_c fv = \int_\tau fv$. Thus define $\int_A f = \int_c fv$.

Proof. (King Lee). Let $c = \sigma_1 + \ldots + \sigma_r$ and $\tau = \gamma_1 + \ldots + \gamma_s$, and throughout this proof let $1 \leq i \leq r$ and $1 \leq j \leq s$. If $A_{ij} = |\sigma_i| \cap |\gamma_j|$, let $B_{ij} = (\sigma_i)^{-1}(A_{ij})$ and $C_{ij} = (\gamma_j)^{-1}(A_{ij})$. Then $\gamma_j^{-1} \circ \sigma_i$ is a diffeo of B_{ij} onto C_{ij} and

$$\int_{B_{ij}} (\sigma_i)^* fv = \int_{B_{ij}} (\sigma_i)^*(\gamma_j \circ \gamma_j^{-1})^* fv = \int_{B_{ij}} (\gamma_j^{-1} \circ \sigma_i)^*(\gamma_j)^* fv = \int_{C_{ij}} (\gamma_j)^* fv.$$

Hence, $\int_c fv = \Sigma_i \int_{\sigma_i} fv = \Sigma_{i,j} \int_{B_{ij}} (\sigma_i)^* fv = \Sigma_{i,j} \int_{C_{ij}} (\gamma_j)^* fv = \int_\tau fv.$//

If M is a compact oriented n-manifold, then M is a fundamental set (this is hard; see Cairns). Thus if M is a compact oriented Riemannian manifold and f is a C^∞ real valued function on M, then $\int_M f$ is well-defined. To handle the non-compact case, define the *support of a function* f to be the set S_f that is the closure of the set $[p$ in M: $f(p) \neq 0]$. Since any compact set of M is contained in a fundamental set (a non-trivial remark), if M is oriented and Riemannian, f is C^∞ with compact support, and $S_f \subset$ fundamental set A, then $\int_M f = \int_A f$ is well-defined (independent of A).

The *area, volume,* or *measure* (depending on the appropriate dimension) of a fundamental set A is the number $\int_A f$, where $f \equiv 1$ on M. For a deeper study of integration theory on manifolds see the book of Whitney[2].

Problems. Let M be a C^∞ n-manifold and let U be an open subset of M.

64. If X and Y are in $T^{1,0}(M)$, f and g in $C^\infty(M, R)$, and w in $T^{0,1}(M)$, show $L_{fX}w = w(X)df + f(L_X w)$, $L_{fX}Y = f(L_X Y) - df(Y)X$, $L_{fX}g = fL_X g$, and $\Delta(fw) = f\Delta w + w \otimes df$. Thus L and Δ are not tensors.

65. If X is a C^∞ vector field on U, m in U, $Z_1, \ldots, Z_n$ a base of M_m with dual base $w_1, \ldots, w_n$, show $(\text{div } X)_m = \Sigma_{i=1}^n w_i(D_{Z_i} X)$. Show that the divergence of a C^∞ field on R^3 agrees with the

advanced calculus definition.

66. Let A be in $T^{1,1}(U)$, let $Z_1,\dots,Z_k$ be a C^∞ base field on U, and let $w_1,\dots,w_k$ be the dual base on U. Show $D_X w_j = -\Sigma_k w_j(D_X Z_k)w_k$ and $\Sigma_j[A(D_X w_j, Z_j) + A(w_j, D_X Z_j)] = 0$.

67. Let M be Riemannian, let $X_1,\dots,X_{n-1}$, T be an orthonormal base, and let P_i be the plane section spanned by X_i and T. Show Ric $(T,T) = \Sigma_{i=1}^{n-1}K(P_i)$.

68. Prove formulas (5), (7), (13), and (14).

69. If D has zero torsion, show $dw(X, Y) = (D_X w)(Y) - (D_Y w)(X)$.

70. If M is Riemannian and $G(X, Y) = \langle X, Y\rangle$, show that a connexion D is metric preserving iff $\Delta G = 0$. Given arbitrary A in $T^{0,3}(M)$ and B in $T^{1,2}(M)$ with $A(X, Y, Z) = A(Y, X, Z)$ and $B(w, X, Y) = -B(w, Y, X)$ for all w, X, Y, Z, show there exists a unique connexion D on M with $\Delta G = A$ and $B(w, X, Y) = w(\mathrm{Tor}_D(X, Y))$.

71. (Poincare lemma) Show every closed p-form on R^n is exact for $p > 0$ as follows: for b in R let g_b: $R^n \to R^{n+1}$ by $g_b(t_1,\dots,t_n) = (t_1,\dots,t_n, b)$, let f: $R^{n+1} \to R$ by $f(t_1,\dots,t_{n+1}) = (t_{n+1}t_1, t_{n+1}t_2,\dots,t_{n+1}t_n)$, let $T = \partial/\partial u_{n+1}$, and for $p > 0$, define the linear map K: $F^p(R^n) \to F^{p-1}(R^n)$ by $K(w) = \int_0^1 (g_b)^* \circ C_T \circ f^*(w)db$, and show $dK + Kd$ equals the identity map on $F^p(R^n)$.

72. Let M be an oriented Riemannian 2-manifold. If σ is an oriented C^∞ curve in M with unit tangent T, let T, N be an orthonormal oriented base along σ and define the *signed geodesic curvature* of σ to be the C^∞ function b with $D_T T = bN$ on σ. If Z, W is an oriented orthonormal parallel base field along σ and $T = (\cos\theta)Z + (\sin\theta)W$, show $b = d\theta/ds = T\theta$ on σ. If x, y is an oriented orthogonal coordinate system on U in M, let $E = \langle X, X\rangle$ and $G = \langle Y, Y\rangle$. If b_1 and b_2 denote the geodesic curvature along the x-coordinate and y-coordinate curves, respectively, show $b_1 = -(1/2E\sqrt{G})(\partial E/\partial y)$, $b_2 = (1/2G\sqrt{E})(\partial G/\partial x)$ and $K = (EG)^{-1/2}[\partial(b_1\sqrt{E})/\partial y - \partial(b_2\sqrt{G})/\partial x]$. Show the y-curve are geodesics (with y as parameter) iff G is constant.

73. If M is Riemannian, (ϕ, U) is a coordinate pair, $x_i = u_i \circ \phi$, $g_{ij} = \langle X_i, X_j \rangle$ where $X_i = \partial/\partial x_i$, $g = \det(g_{ij})$, f is in $C^\infty(M, R)$, and A is a fundamental set with $A \subset U$, show $\int_A f = \int_{\phi(A)} (f \circ \phi^{-1})\sqrt{g \circ \phi^{-1}}\, du_1 du_2 \ldots du_n$.

74. Let M be a surface in R^3 with sphere map η. For m in M let $A(r)$ be the area of $B(m, r)$, the ball about m of radius r and let $A_\eta(r)$ be the area of $\eta(B(m, r))$. Show $K(m) = \lim [A_\eta(r)/A(r)]$ as $r \to 0$.

8. Gauss-Bonnet Theory and Rigidity

In this chapter, M will denote a connected oriented Riemannian n-manifold.

Section 8.1. Gauss-Bonnet formula.

In this section, let $n = 2$, let A be a fundamental set in M, and let c be a fundamental 2-chain with $|c| = A$. The oriented curve $\gamma = \partial c$ is called the *bounding curve* of A. A *vertex of* c is a point in M that is the image of a vertex in I^2 under a 2-cube in c. A *face of* c is the support of a 2-cube in c. An *edge of* c is the face of a 1-cube in $\partial\sigma$ for some 2-cube σ in c. A *boundary edge of* c is an edge that is in γ. A *corner point* of γ is a vertex of c belonging to exactly two boundary edges. At a corner point p of γ, let $T_i(p)$ (the "tangent in") and $T_0(p)$ (the "tangent out") be the unit tangents at p of the 1-cubes in γ, defined by the orientation, going "into" and "out from" p, respectively. The *exterior corner angle* $\alpha(p)$ is the angle such that $\cos \alpha(p) = \langle T_i(p), T_0(p) \rangle$ and $0<\alpha<\pi$ or $-\pi<\alpha<0$ according as T_i, T_0 is a positively or negatively oriented base. If $T_0 = T_i$, then $\alpha = 0$, and if $T_0 = -T_i$ then $\alpha = -\pi$ (see Fig. 8.1).

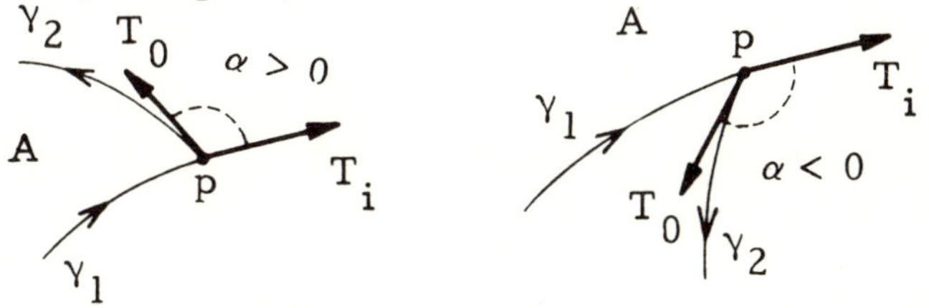

Fig. 8.1 Corner Angles

In the proof of the Gauss-Bonnet formula that follows, the differential geometry involved is simple. The crux of the theorem is the Hopf Umlaufsatz (see discussion after proof). As usual, a *simple closed curve* is a homeomorphic image of the circle S^1 in R^2.

THEOREM (Gauss-Bonnet formula). *Let A be contained in a coordinate domain U of M, let the bounding curve γ of A be a simple closed curve, and let $a_1,\ldots,a_r$ be the exterior corner angles of γ. Then*

$$(1) \qquad \int_\gamma k = 2\pi - \Sigma_{j=1}^r a_j - \int_A K$$

where k is the signed geodesic curvature function on γ and K is the Riemannian (Gaussian) curvature function on A.

Proof. Let e_1, e_2 be a C^∞ positively oriented base field on U. Let $\gamma_1,\ldots,\gamma_r$ be the C^∞ pieces of γ with each γ_j parameterized by arc length on the interval $[s_j, s_{j+1}]$, $\gamma_j(s_{j+1}) = \gamma_{j+1}(s_{j+1})$ for $j = 1,\ldots,r-1$, while $\gamma_r(s_{r+1}) = \gamma_1(s_1)$, and a_j the exterior corner angle at $\gamma(s_j)$. Let T be the unit tangent to γ. By making a constant rotation of e_1, e_2, if necessary, we may assume $T(s_1+) = e_1$. Define $\zeta(s)$ on $[s_1, s_2]$ so ζ is C^∞, $\zeta(s_1^+) = 0$, and $T = (\cos\zeta)e_1 + (\sin\zeta)e_2$. This ζ is well-defined, since we have given its initial value and it is C^∞, since locally it is given by $\zeta(s) = \cos^{-1}\langle T(s), e_1(s)\rangle$ for a proper branch of the inverse cosine. Thus we obtain $\zeta(s_2^-)$. Let $\zeta(s_2^+) = \zeta(s_2^-) + a_2$ and extend ζ to $[s_2, s_3]$ so ζ is C^∞ and $T = (\cos\zeta)e_1 + (\sin\zeta)e_2$, as before. Continuing this process, we extend ζ to $[s_1, s_{r+1}]$ with ζ in C^∞ at all interior points except s_i where it has a jump precisely equal to a_i for $i = 2,\ldots,r$. Since γ is a simple closed curve, we use the Hopf Umlaufsatz to obtain $\zeta(s_{r+1}^-) + a_1 = \zeta(s_1^+) + 2\pi$. We include a schematic diagram (Fig. 8.2):

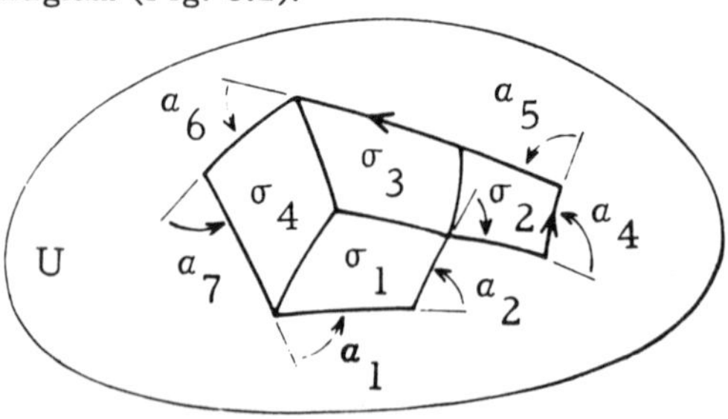

Fig. 8.2 Fundamental Set

On each C^∞ piece of γ we have the positively oriented orthonormal base field, T, N, and the *signed geodesic* cirvature k is defined by $D_T T = kN$. In terms of ζ, $T = (\cos\zeta)e_1 + (\sin\zeta)e_2$, while $N = (-\sin\zeta)e_1 + (\cos\zeta)e_2$.

Let w_1, w_2 be the dual 1-forms to the base e_1, e_2, and let $w_{12} = -w_{21}$ be the corresponding connexion 1-form on U (note $w_{11} = w_{22} = 0$ for the Riemannian connexion D). Thus $v = w_1 \wedge w_2$ is the volume element on U. Moreover, by the Cartan structural equations, $dw_{12} = R_{12}$, and $K = \langle R(e_1, e_2)e_2, e_1\rangle = \langle \Sigma_{i=1}^2 R_{i2}(e_1, e_2)e_i, e_1\rangle = R_{12}(e_1, e_2)$, thus $R_{12} = Kw_1 \wedge w_2$.

Since $k = \langle D_T T, N\rangle$ and $D_T T = (T\zeta)N + (\cos\zeta)w_{21}(T)e_2 + (\sin\zeta)w_{12}(T)e_1$, then

(2) $\quad k = (T\zeta) - w_{12}(T),$

which is a *Cartan formula for the geodesic curvature*. Then

$$\int_\gamma k = \Sigma_{j=1}^r \int_{s_j}^{s_{j+1}} \frac{d\zeta}{ds}\,ds - \int_{\partial c} w_{12} = \Sigma_{j=1}^r [\zeta(s_{j+1}^-) - \zeta(s_j^+)] - \int_c dw_{12}$$

$$= 2\pi - \Sigma_{j=1}^r \alpha_j - \int_A K,$$

where we use Stokes' theorem for the second equality.//

The Gauss-Bonnet formula almost proves the Hopf Umlaufsatz (see Hopf[1]), which states if γ is a simple closed smooth (C^1) curve in R^2, then $\int_\gamma k = \pm 2\pi$, depending on the orientation of γ. We need the topological result that γ disconnects the plane into two components and the map γ may be extended to a homeomorphism of the interior of the disc $B(0, 1)$, which then maps onto a set A, which is fundamental and has γ as bounding curve. Then letting $e_1 = i$, $e_2 = j$ (advanced calculus notation), we have $w_{12} = 0$, $K = 0$, and all $\alpha_i = 0$, so $\int_\gamma k = 2\pi$ if γ positively oriented. The reader may also be interested in the papers of H. Whitney[1], J. S. Griffin, and C. J. Titus.

The Gauss-Bonnet formula was first proved by Bonnet in 1848. Somewhat earlier Gauss had proved the following result on geodesic triangles.

THEOREM (Gauss). *Let A be a fundamental set of M bounded by three (non-closed) geodesics, i.e., A is a* geodesic triangle, *and let β_1, β_2, and β_3 be the interior angles at the corners. Then $\int_A K = \beta_1 + \beta_2 + \beta_3 - \pi$, and this number is called the* excess of the triangle.

Proof. The Gauss-Bonnet formula is applicable. Since $k = 0$ and $a_i = \pi - \beta_i$, we have $0 = 2\pi - \Sigma_1^3(\pi - \beta_j) - \int_A K$.//

Corollary. Let B be the sum of the interior angles of a geodesic triangle A on M. Then B is $>\pi$, $=\pi$, or $<\pi$, according as $K > 0$, $= 0$, or < 0 on A. If K is constant and not zero on A, then the area of A equals the excess of A divided by K.

We obtain some simple applications of the Gauss-Bonnet formula by applying it to the cases when M is diffeo to the sphere or the torus. In the former case $\int_M K = 4\pi$, and in the latter case $\int_M K = 0$. These are special cases of the Gauss-Bonnet theorem which we prove later in this section. We sketch the proofs of these facts.

When M is diffeo to S^2, we let γ be the image of the equator (under the diffeo), A_1 the image of the "northern" hemisphere, and A_2 the image of the "southern" hemisphere (see Fig. 8.3). Supposing γ to be the bounding curve of A_1, we have

$$\int_\gamma k = 2\pi - \int_{A_1} K \quad \text{and} \quad \int_{-\gamma} k = -\int_\gamma k = 2\pi - \int_{A_2} K.$$

Hence $$\int_M K = \int_{A_1} K + \int_{A_2} K = 4\pi.$$

When M is diffeo to the torus, let A_1 be the image of the "top half" and A_2 the image of the "bottom half" of the torus so A_1 and A_2 are bounded and separated by the image γ of the "inside" and "outside" curve on the torus (see Fig. 8.3). Again letting γ be the bounding curve of A_1, connecting and closing γ via a cut curve β (see Fig. 8.3), and taking a limit, we obtain $\int_\gamma k = 2\pi - 2\pi - \int_{A_1} K$ and $-\int_\gamma k = -\int_{A_2} K$ so $\int_M K = 0$.

Our next task is to free the Gauss-Bonnet formula from the special neighborhood U. The proof follows the notes of Samelson. Define the *Euler characteristic*, $\mathcal{X}_c(A)$, *of A with respect to* c by $\mathcal{X}_c(A) = V - E + F$, where V is the number of vertices of c, E the number of

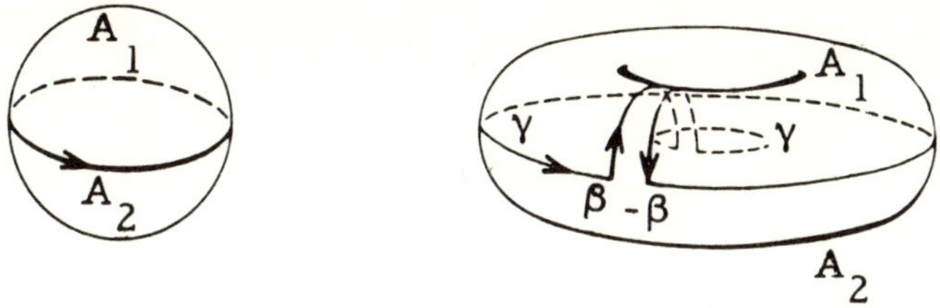

Fig. 8.3 Sphere and Torus

edges, and F the number of faces.

THEOREM. *Let A be a fundamental set on M, let the bounding curve γ of A be a finite disjoint union of simple closed curves, and let $a_1,\dots,a_r$ be the exterior corner angles of γ. Then*

$$(3) \qquad \int_\gamma k = 2\pi\mathcal{X}_c(A) - \Sigma_{i=1}^r a_i - \int_A K.$$

This expression proves $\mathcal{X}_c(A)$ is independent of c, so define $\mathcal{X}(A) = \mathcal{X}_c(A)$ to be the Euler characteristic *of A and drop the subscript c in the above formula.*

Proof. Let $c = \sigma_1+\dots+\sigma_F$ and note from the definition of a fundamental 2-chain we may apply the Gauss-Bonnet formula to each set $|\sigma_j|$ (for σ_j defines a coordinate neighborhood of $|\sigma_j|$). Let $a_1^j,\dots,a_4^j$ denote the four exterior angles for σ_j. Then

$$\int_\gamma k = \Sigma_{j=1}^F \int_{\partial\sigma_j} k = \Sigma_{j=1}^F (2\pi - \Sigma_{i=1}^4 a_i^j) - \Sigma_{j=1}^F \int_{\sigma_j} K$$

or

$$\int_\gamma k = 2\pi F - \Sigma_{j=1}^F \Sigma_{i=1}^4 a_i^j - \int_A K.$$

Thus the problem is one of bookkeeping with the term Σa_i^j.

Let β_i^j be the interior angle corresponding to each a_i^j, thus $\beta_i^j = \pi - a_i^j$, and let $\beta_s = \pi - a_s$ be the interior angles at the corners of γ. In the following we sum over $i = 1,\dots,4$ and $j = 1,\dots,F$. The sum

$$\Sigma_{ij}\beta_i^j = 2\pi(V - r) + \Sigma_1^r\beta_s = 2\pi(V - r) + \Sigma_1^r(\pi - a_s)$$

$$= 2\pi V - \pi r - \Sigma_1^r a_s,$$

since r is the number of verticies of c on γ (as well as the number of angles and edges on γ) so $(V - r)$ is the number of vertices interior to A, each of which contributes 2π to the total sum.

We now show $rF = (2E - r)$, which is the number of terms in the sum $\Sigma_{ij}\beta_i^j$. This is done by assigning to each β_i^j an edge, namely, its "starting" edge, which is well-defined by the orientation. More precisely, if T and T' are the unit vectors at the vertex of β_i^j which are tangent to the edge curves of β_i^j, then T and T' are independent, since c is a fundamental chain, so $(\sigma_j)_*$ is non-singular on its domain (which is slightly larger than I^2). Thus T is the "starting" edge of β_i^j if and only if T, T' is a positively oriented bases (see Fig. 8.4).

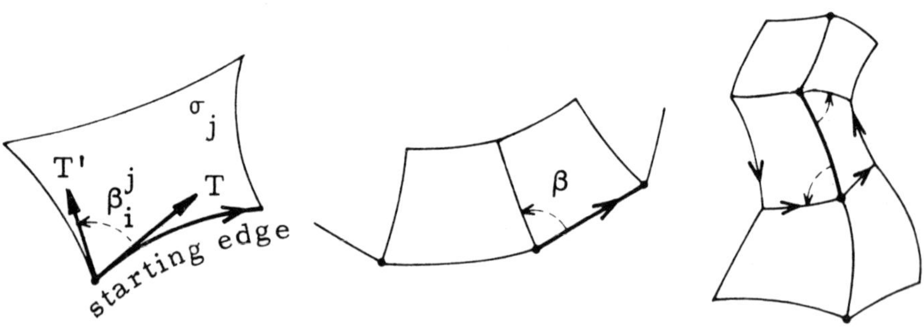

Fig. 8.4 Starting Edges

Then each edge on the boundary γ belongs to exactly one β_i^j, while each edge not on the boundary belongs to exactly two β_i^j. Thus $rF = r + 2(E - r)$, since r is the number of edges on the boundary.

Finally,

$$\Sigma_{ij}a_i^j = \Sigma_{ij}(\pi - \beta_i^j) = \pi(2E - r) - 2\pi V + \pi r + \Sigma_1^r a_s = 2\pi(E - V) +$$

$$\Sigma_1^r a_s.$$

Hence,

$$\int_\gamma k = 2\pi(F - E + V) - \Sigma_1^r a_s - \int_A K.//$$

GAUSS-BONNET THEOREM. *Let M be a compact connected oriented Riemannian 2-manifold with Riemannian (Gaussian) curvature function K. Then $\int_M K = 2\pi\mathfrak{X}(M)$.*

Proof. We apply the preceding theorem to a fundamental chain on M which will have no boundary and no exterior angles.//

The above theorem is an important example of a theorem relating differential geometry and topology. The Euler characteristic is a topological invariant which does not depend on either the differentiable structure or the Riemannian structure on M. The theorem may be used to prove many "negative" statements: for example, there does not exist a Riemannian metric on the torus with $K > 0$ everywhere (nor does there exist one with $K < 0$ everywhere) since $\mathfrak{X}(M) = 0$ (which we computed above for the induced Riemannian metric). The theorem has been generalized for dimensions greater than two and provides one of the first successes of the global theory of fiber bundles.

Section 8.2 Index theorem.

This section is also based on the notes of Samelson. Let $n = 2$ and let W be a C^∞ vector field on M. If $W_m = 0$, then m is a *singularity* of W. Assuming W has only isolated singularities, we define the *index of W at m, $J(W, m)$,* as follows.

Let U be a coordinate domain, with coordinate radius $b > 0$, about m with $W \neq 0$ on $U - [m]$. Assume the coordinate map is orientation preserving, and let σ_r be the oriented coordinate circle of radius r about m with $0 < r < b$ and σ_r defined on $[0, 1]$. Let X be a unit vector field on U. Since W does not vanish on σ_r, by using the proper inverse cosine function one obtains a C^∞ function θ on $[0, 1]$ with $\langle W(s), X(s)\rangle = |W(s)| \cos\theta(s)$ on $[0, 1]$. Let

$$(4) \qquad 2\pi J_X(W, m, r,) = \theta(1) - \theta(0).$$

For $0 < r < b$, $J_X(W, m, r)$ is a continuous integer-valued function, and hence yields a constant $J_X(W, m)$. If m is not a singular point, then for small $r > 0$, θ is close to the constant $\cos^{-1}(\langle W_m, X_m\rangle / |W_m|) (\bmod 2\pi)$; hence $\theta(1) = \theta(0)$ and $J_X(W, m) = 0$. If Y is another unit vector field on U, then

$$J_X(W, m) = J_Y(W, m) + J_X(Y, m) = J_Y(W, m),$$

since Y has no singularities. Thus $J_X(W, m)$ is independent of X. An analogous argument shows $J(W, m)$ can be computed by using any simple closed C^∞ curve σ about m with σ in U, and thus $J(W, m)$ is an integer depending only on W and m (see Fig. 8·5).

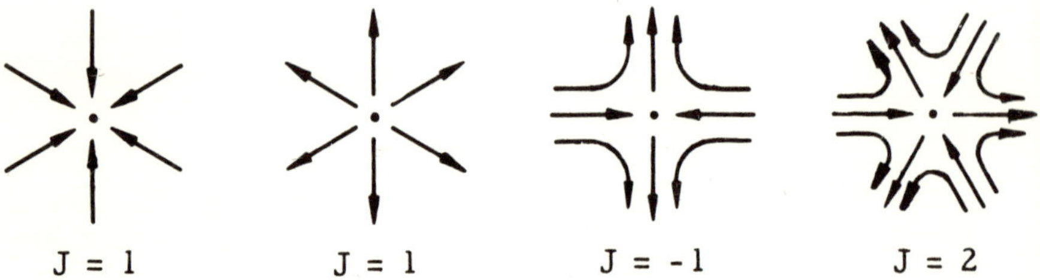

Fig. 8.5 Examples of $J(W, m)$

If W has only a finite number of singularities, define the *index of* W, $J(W)$, by $J(W) = \Sigma_m J(W, m)$.

INDEX THEOREM. *If M is a compact connected oriented Riemannian 2-manifold and W is a C^∞ vector field on M with a finite number of singularities, then the index of W equals the Euler characteristic of M.*

Proof. Take an oriented fundamental chain $c = \sigma_1 + \ldots + \sigma_r$ with at most one singularity m_i of W in the interior of each $|\sigma_i|$. Let γ_i be the bounding curve of σ_i and define functions θ_i, ζ_i and ξ_i on the domain of γ_i so $\theta = \zeta_i + \xi_i$, θ_i is an angle between W and e_i, ζ_i is an angle between the tangent T_i of γ_i and e_1, and ξ_i is an angle between W and T_i. The functions θ_i, ζ_i, and ξ_i will be piece-wise C^∞ and θ_i is continuous. By integrating over the pieces of γ_i we obtain

$$2\pi J(W, m_i) = \int_{\gamma_i} \frac{d\theta_i}{ds} = \int_{\gamma_i} \frac{d\zeta_i}{ds} + \int_{\gamma_i} \frac{d\zeta_i}{ds},$$

$$= \int_{\gamma_i} k + \int_{\sigma_i} K + \int_{\gamma_i} \frac{d\xi_i}{ds}.$$

Adding over the 2-cubes in c gives

(5) $\quad 2\pi J(W) = \int_M K$

since the integrals over the bounding curves cancel one another. By the Gauss-Bonnet theorem, $J(W) = \mathcal{X}(M)$.//

Omitting the last line of the proof, we note $2\pi J(W) = \int_M K$ implies $J(W)$ is independent of W as long as W has only a finite number of singularities. Then for any oriented fundamental chain c we can define a particular W which has a singularity for each face, edge, and vertex with index 1, -1, and 1, respectively. We indicate in Fig. 8.6 how W is defined on each 2-cube. Actually W would be precisely defined by defining a field on a neighborhood of I^2 and carrying this to each $|\sigma_i|$ via the map σ_i.

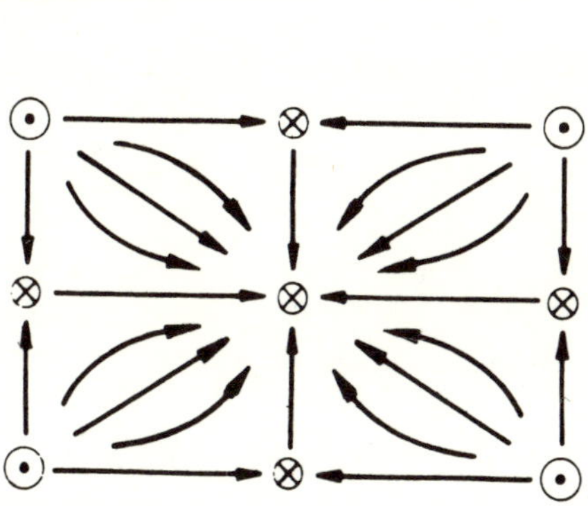

Fig. 8.6 The Canonical Vector Field on a 2-cube

Thus W is defined by "going out from each vertex and in to the center of each face." From Fig. 8.6 we see $J(W) = V - E + F = \mathcal{X}_c(M)$. Thus we again prove $\mathcal{X}_c(M)$ is independent of c, and $2\pi\mathcal{X}(M) = 2\pi J(W) = \int_M K$ reproves the Gauss-Bonnet theorem.

Corollary. If M is a manifold as described in the theorem and there exists a non-vanishing C^∞ vector field on M, then $\mathfrak{X}(M) = 0$. Thus any surface that is diffeomorphic to the 2-sphere has no non-vanishing C^∞ vector fields.

Actually, a differentiable manifold (any dimension) admits a non-zero continuous vector field if and only if its Euler characteristic is zero (see Steenrod, p. 203, and Alexandroff and Hopf, p. 549).

Section 8.3. Gauss-Bonnet form.

In the proof of the Gauss-Bonnet formula we found that R_{12} is a local representation of the global form Kv on an oriented Riemannian 2-manifold M. One might ask if there are other global forms obtainable in this way, or if there is an analogous form on an n-manifold. We answer these questions now.

Let $e_1,\ldots,e_n$ and $f_1,\ldots,f_n$ be two set of positively oriented orthonormal C^∞ base fields on an open set U in M, and let $f_j = \Sigma_{i=1}^n b_{ij} e_i$ define C^∞ functions b_{ij} on U. Notice that determinant $(b_{ij}) = 1$ and $(b_{ij})^{-1} = (b_{ji})$ since (b_{ij}) is orthogonal. We let R_{ij} and $\overline{R}_{ij}$ denote the local curvature forms associated with e_i's and f_i's, respectively, thus $R(X, Y)e_j = \Sigma R_{ij}(X, Y)e_i$. Then for m in U, X and Y in M_m, we have

$$(6) \quad R_{ij}(X, Y) = \langle R(X, Y)e_j,\ e_i\rangle = \langle R(X, Y)(\Sigma_r b_{jr} f_r),\ \Sigma_s b_{is} f_s\rangle$$

$$= \Sigma_{r,s} b_{jr} b_{is} \langle R(X, Y)f_r,\ f_s\rangle = \Sigma_{r,s} b_{is} \overline{R}_{sr}(X, Y) b_{jr}.$$

Thus $R_{ij} = \Sigma_{r,s} b_{is} \overline{R}_{sr} b_{jr}$ relates the local curvature forms of the two bases on U.

If n is even, we define an n-form Q on U by

$$(7) \quad Q = \Sigma(-1)^\pi R_{\pi(1)\pi(2)} \wedge R_{\pi(3)\pi(4)} \wedge \cdots \wedge R_{\pi(n-1)\pi(n)}$$

where we sum over all permutations π in P_n, the group of permutations on the set $[1, 2,\ldots,n]$. The representation of Q in terms of the forms $\overline{R}_{ij}$ is,

$$Q = \Sigma_{r_i=1}^n \Sigma(-1)^\pi b_{\pi(1)r_1} \overline{R}_{r_1 r_2} b_{\pi(2)r_2} b_{\pi(3)r_3} \overline{R}_{r_3 r_4} b_{\pi(4)r_4} \cdots$$

$$= (\det b_{ij})\Sigma(-1)^{\pi}\bar{R}_{\pi(1)\pi(2)} \wedge \bar{R}_{\pi(3)\pi(4)} \wedge \cdots \wedge \bar{R}_{\pi(n-1)\pi(n)}$$

Since $(\det b_{ij}) = 1$, Q is independent of the particular base field used to define it; thus Q defines a global n-form on M which is called the *Gauss-Bonnet form.* Note if $n = 2$, then locally $Q = R_{12} - R_{21} = 2R_{12} = 2Kv$.

THEOREM (Generalized Gauss-Bonnet). *If M is an even dimensional ($n = 2k$) compact connected oriented Riemannian manifold, then $\int_M Q = 2^n\pi^k(k!)\mathfrak{X}(M)$.*

For a proof see Chern[1]. Other pertinent references are Hopf[1], Allendoerfer[1], Allendoerfer and Weil, Fenchel[1], Chern[3], and Allendoerfer[2].

Let M be as in the theorem and assume further that M is a hypersurface in R^{n+1} with unit normal field N. Using the notation from section 4, $R_{ij} = -w_{in+1} \wedge w_{n+1j} = w_{in+1} \wedge w_{jn+1}$ and $L(X) = \Sigma_{i=1}^{n} w_{in+1}(X)e_i = \eta_*(X)$ where η is the sphere map induced by the normal N (section 2.2). Thus $\eta^* w_i = w_{in+1}$ and

$$Q = \Sigma(-1)^{\pi} w_{\pi(1)n+1} \wedge \cdots \wedge w_{\pi(n-1)n+1} \wedge w_{\pi(n)n+1}$$

$$= n!\, w_{1n+1} \wedge w_{2n+1} \wedge \cdots \wedge w_{nn+1} = n!\, \eta^*(v_S),$$

where v_S is the volume element of the unit sphere S^n oriented by its outer normal, and we assume N_m is parallel to the outer normal at $\eta(m)$. Integrating, $\int_M Q = (n!)\int_M \eta^*(V_S)$. If $n = 2$, then $\int_M Q = 2\int_M Kv = 4\pi\mathfrak{X}(M) = 2\int_M \eta^*(v_S)$, thus $\int_M \eta^*(V_S) = (V_2)\,\mathfrak{X}(M)/2$ where V_2 is the "volume" of the unit 2-sphere. This is the *Hopf Index theorem* for dimension 2.

In the general case (M^n imbedded in R^{n+1} as above). we let X be any unit vector field of R^{n+1} that is C^∞ on M, and we define the *index of X on M*, $\mathfrak{I}(X)$, by

(8) $\quad \mathfrak{I}(X) = (1/V_n)\int_M \eta_X{}^*(v_S)$

where V_n is the "volume" of the unit n-sphere in R^{n+1}, and η_X is the C^∞ map of M into S^n induced by the vector field X.

HOPF INDEX THEOREM. *If M^n is an even dimensional compact connected submanifold of R^{n+1}, then twice the index of the normal field N on M is the Euler characteristic of M, or $2\mathcal{I}(N) = \mathcal{X}(M)$.*

Proof. Assuming the Gauss-Bonnet theorem and letting $n = 2k$, $2\mathcal{I}(N) = 2\int_M Q/(V_n n!) = 2^{n+1}\pi^k k!\mathcal{X}(M)/(V_n n!) = \mathcal{X}(M)$, since $V_n = 2^{n+1}\pi^k k!/n!$ (see problem 75.).//

Section 8.4. Characteristic forms.

A general reference for this section is Chern[1] with related treatments in Adler and H. Cartan. The "wedge" product symbol between forms will be omitted in this section.

For $k > 1$, define local forms

$$Q_k = \Sigma^n_{i_j=1} R_{i_1 i_2} R_{i_2 i_3} R_{i_3 i_4} \ldots R_{i_k i_1},$$

where the R_{ij} belong to a local positively oriented orthonormal base field $e_1, \ldots, e_n$. As above (equation (7)), one shows Q_k is independent of this particular base field and thus Q_k is a global $2k$-form on M. Moreover $dQ_k = 0$, i.e., each Q_k is a closed $2k$-form. To prove this, use

$$dR_{ij} = \Sigma^n_{r=1}(R_{ir}w_{rj} - R_{jr}w_{ri}),$$

which follows from the second structural equation (section 5.2). Then,

$$dQ_k = \Sigma[(dR_{i_1 i_2})R_{i_2 i_3} \ldots R_{i_k i_1} + R_{i_1 i_2}(dR_{i_2 i_3})R_{i_3 i_4} \ldots R_{i_k i_1} + \ldots].$$

Consider one of the sums (all indices are summed from 1 to n), $A = \Sigma R_{i_1 r} w_{r i_2} R_{i_2 i_3} \ldots R_{i_k i_1}$. If k is even, the products in A are formed from an odd number of forms that are skew-symmetric in their indices; hence switching all the indices changes the sign, and adding, one gets $A = -A$ so $A = 0$. If k is odd, the argument just used shows $Q_k = 0$.

Proposition. For even k, the forms Q_k define global closed $2k$-forms on M. For odd k, $Q_k = 0$.

Let W_D denote the subalgebra of the Cartan differential algebra F(or $F(M)$) which is generated over the real field by the forms Q_k for $k = 2, 4, \ldots [n/2]$, and call W_D the *algebra of characteristic forms for the connexion D*. Elements in W_D are called *characteristic forms*, and they are closed forms since the generators are all closed. By going to the differential cohomology we can free ourselves of the connexion D which we now do.

Let F^p denote the module of C^∞ p-forms on M. Let Z^p denote the *closed* forms in F^p, thus $Z^p = [\alpha \text{ in } F^p\text{: } d\alpha = 0]$; and let B^p denote the *exact* forms in F^p, so $B^p = [\alpha \text{ in } F^p\text{: there is } \beta \text{ in } F^{p-1} \text{ with } d\beta = \alpha]$. Since $d^2 = 0$, $B^p \subset Z^p$; hence let $H^p = Z^p/B^p$ and call H^p the *p-dimensional differential cohomology group of M*. If α in Z^p, denote its image in H^p by $\overline{\alpha}$; hence $\overline{\alpha}$ is the coset $\alpha + B^p$ which is called a *(differential) cohomology class* on M. Let $H^* = \sum_{p=0}^{n} H^p$ (direct sum) and notice the multiplication in F carries over to H^*.

Thus $\overline{W}_D$ defines a set of classes called *(differential) characteristic cohomology classes*, and this set we show is independent of D (the Riemannian structure) and depends only on the manifold M. It is customary to speak of $\overline{W}_D$ as the image of the *Wiel homomorphism*. This we explain.

Let $g\ell\,(n, R)$ be the set of n by n matrices over the real field R. Our notation is the customary one for this set when it is thought of as the Lie algebra of the general linear group $GL(n, R)$. If $A = (a_{ij})$ in $g\ell(n, R)$ we let $u_{ij}(A) = a_{ij}$. Then a *polynomial function P* on $g\ell(n, R)$ is a polynomial in the functions $u_{11}, u_{12}, \ldots, u_{nn}$; for example, $P(A) =$ determinant (A) is a polynomial function. An *invariant polynomial P* on $g\ell(n, R)$ is a polynomial function P such that $P(BAB^{-1}) = P(A)$ for all non-singular orthogonal matrices B. Referring to the way we define the characteristic forms Q_k, we see that every *invariant polynomial P can be used to define a global differential form Q on M* by using the curvature forms from a Riemannian connexion D on M and letting $Q = P(R_{11}, R_{12}, \ldots, R_{nn})$. Let us use $\mathfrak{W}_D$ for this map, so $Q = \mathfrak{W}_D(P)$. Letting $\mathfrak{I}$ denote the set of invariant polynomials on $g\ell(n, R)$, we then claim to have a homomorphism $\mathfrak{W}_D\colon \mathfrak{I} \to F(M)$ with $\mathfrak{W}_D(\mathfrak{I}) = W_D(M) = W_D$. This is the *Weil homomorphism*.

THEOREM. *The Weil homomorphism is well-defined from the set of invariant polynomials on $g\ell(n, R)$ onto the set of characteristic* *differential forms on M; moreover, the Weil homomorphism is independent of the connexion at the cohomology level, i.e., $\overline{W}_{D_1} = \overline{W}_{D_2}$ for two Riemannian connexions D_1 and D_2.*

Proof. Let $f_A(\lambda)$ denote the characteristic polynomial of a matrix A and define polynomials $E_r(A)$ to be the coefficients in $f_A(\lambda)$, thus $f_A(\lambda) = \det(\lambda I - A) = \lambda^n + E_{n-1}(A)\lambda^{n-1}+\ldots+E_0(A)$. From linear algebra we know $E_r(A)$ are *invariant polynomials* on $g\ell(n, R)$; moreover, they generate the ring of invariant polynomials. In terms of the characteristic roots of A, $E_r(A)$ is the rth elementary symmetric function of these roots, i.e., $E_1(A) = a_1+\ldots+a_n$, $E_2(A) = \Sigma_{i<j}a_ia_j$, etc. By Newton's theorem on symmetric functions, the functions $E_r(A)$ are expressible as polynomials in the functions $P_r(A)$, where $P_r(A) = (a_1)^r + (a_2)^r +\ldots+(a_n)^r$. But $P_r(A)$ is the trace of A^r, and we can write this trace in terms of the elements of A by

$$P_r(A) = \Sigma a_{i_1 i_2} a_{i_2 i_3} \ldots a_{i_r i_1}$$

summing over all $i_j = 1,\ldots,n$. Hence $\mathfrak{W}_D(\mathfrak{I})$ is generated by the forms Q_k, $\mathfrak{W}_D$ is well-defined, and $\mathfrak{W}_D(\mathfrak{I}) = W_D$.

To show $\overline{W}_D$ is independent of the Riemannian connexion, we take two such connexions D_1 and D_0, let $Q_k^i = W_{D_i}(P_k)$ for $i = 0, 1$, and show $Q_k^1 - Q_k^0 = dG$, where G in $F^{2k-1}(M)$. Thus $\overline{Q}_k^1 = \overline{Q}_k^0$, which implies $\overline{W}_{D_1} = \overline{W}_{D_0}$.

Let $<X, Y>_1$ and $<X, Y>_0$ be the Riemannian metrics associated with D_1 and D_0, respectively, and for $0 \le t \le 1$ define $<X, Y>_t = t<X, Y>_1 + (1-t)<X, Y>_0$. Then $<X, Y>_t$ is a Riemannian metric for each t, and its Riemannian connexion D_t is given by $D_t = tD_1 + (1-t)D_0$. This can be shown easily by verifying that D_t has zero torsion and preserves the metric $<X, Y>_t$. For any base field $e_1,\ldots,e_n$ on an open set U of M let w_{ij}^t and R_{ij}^t be the connexion and curvature forms associated with D_t. Then $(D_t)_X e_j = \Sigma_i w_{ij}^t(X)e_i = t\Sigma_i w_{ij}^1(X)e_i + (1-t)\Sigma_i w_{ij}^0(X)e_i$, so $w_{ij}^t = tw_{ij}^1 - (1-t)w_{ij}^0$. From the second Cartan structural equation we obtain

$$R_{ij}^t = tR_{ij}^1 + (1-t)R_{ij}^0 + t(t-1)\Sigma_k \theta_{ik}\theta_{kj},$$

where $\theta_{ij} = w^1_{ij} - w^0_{ij}$. The 1-forms θ_{ij} are the local forms belonging to the difference tensor $B(X, Y) = (D_1)_X Y - (D_0)_X Y$, i.e., $B(X, e_j) = \Sigma_i \theta_{ij}(X) e_i$. Since B is a tensor, if $f_1, \ldots, f_n$ is another base field on U with $f_j = \Sigma_i b_{ij} e_i$ and $B(X, f_j) = \Sigma_i \overline{\theta}_{ij}(X) f_i$, then $\theta_{ij}(X) = \Sigma_{r,s} b_{is} \theta_{sr}(X)(b^{-1})_{rj}$.

For each even k and each t, choose $e_1, \ldots, e_n$ to be an orthonormal base field relative to the metric $\langle X, Y\rangle_t$, and define a $(2k - 1)$ form on U by $G^t_k = \Sigma \theta_{i_1 i_2} R^t_{i_2 i_3} R^t_{i_3 i_4} \ldots R^t_{i_k i_1}$, summing over all $i_j = 1, \ldots, n$. Since the θ_{ij} transform exactly like the R_{ij} when changing to another orthonormal base, the forms G^t_k are global forms on M by the argument that was used to show Q_k are global forms. Note θ_{ij} are *not* skew-symmetric.

In an obvious way, define for each t, a $2k$-form $(d/dt)Q^t_k$, i.e.,

$$\frac{d}{dt} Q^t_k = \frac{d}{dt} (\Sigma R^t_{i_1 i_2} R^t_{i_2 i_3} \ldots R^t_{i_k i_1}) = k \Sigma \left(\frac{d}{dt} R^t_{i_1 i_2} \right) R^t_{i_2 i_3} \ldots R^t_{i_k i_1}$$

where $\frac{d}{dt} R^t_{ij} = R^1_{ij} - R^0_{ij} + (2t - 1)\Sigma_k \theta_{ik}\theta_{kj}$. Then $Q^1_k - Q^0_k = \int_0^1 \frac{d}{dt}(Q^t_k)dt$, and we show $kdG^t_k = (d/dt)Q^t_k$; hence $Q^1_k - Q^0_k = dG_k$ where $G_k = k\int_0^1 G^t_k dt$.

To compute dG^t_k, use the second Cartan structural equation to obtain $d\theta_{ij} = R^1_{ij} - R^0_{ij} - \Sigma[\theta_{ik}\theta_{kj} + \theta_{ij} w^0_{kj} + w^0_{ik}\theta_{kj}]$. Also,

$$dR^t_{ij} = \Sigma(R^t_{ik} w^t_{kj} - w^t_{ik} R^t_{kj}) = \Sigma[tR^t_{ik}\theta_{kj} - t\theta_{ik} R^t_{kj} + R^t_{ik} w^0_{kj} - w^0_{ik} R^t_{kj}],$$

since $w^t_{ij} = t\theta_{ij} + w^0_{ij}$. Hence

$$dG^t_k = \Sigma(R^1_{i_1 i_2} - R^0_{i_1 i_2}\ \theta_{i_1 k}\theta_{k i_2} - \theta_{i_1 k} w^0_{k i_2} - w_{i_1 k}\theta_{k i_2}) R^t_{i_2 i_3} \ldots R^t_{i_k i_1}$$

$$- [\Sigma\theta_{i_1 i_2}(tR^t_{i_2 k}\theta_{k i_3} - t\theta_{i_2 k} R^t_{k i_3} + R^t_{i_2 k} w^0_{k i_3} - w^0_{i_2 k} R^t_{k i_3}) R^t_{i_3 i_4} \ldots R^t_{i_k i_1}] \ldots$$

$$- [\Sigma\theta_{i_1 i_2} R^t_{i_2 i_3} \ldots R^t_{i_{k-1} i_k}(tR^t_{i_k j}\theta_{j i_1} - t\theta_{i_k j} R^t_{j i_1} + R^t_{i_k j} w^0_{j i_1} - w^0_{i_k j} R^t_{j i_1}$$

$$= \Sigma[R^1_{i_1 i_2} - R^0_{i_1 i_2} + (2t-1)\Sigma\theta_{i_1 k}\theta_{k i_2}]R^t_{i_2 i_3}\ldots R^t_{i_k i_1} = \frac{1}{k}\frac{d}{dt}Q^t_k.//$$

Section 8.5. *Rigidity problems.*

Two submanifolds of R^n are *congruent* or *symmetric* if there is an isometry of R^n mapping one onto the other that is orientation preserving or reversing, respectively. Let us say a submanifold M of R^n is *rigid* if any submanifold M' that is isometric to M is actually congruent or symmetric to M. Natural questions arise which are called *rigidity problems*. For example, which submanifolds are rigid, or when are two isometric submanifolds congruent or symmetric?

Our principal reference for this section is Chern[1]. The standard procedure in the following theorems is to somehow set up the hypothesis of the fundamental rigidity theorem proved in section 6.5. Given an isometry f between submanifolds, the first fundamental form is preserved by hypothesis, and our task is to show the second fundamental form is preserved, or that f_* commutes with the fundamental linear transformations L.

THEOREM. *If $n \geq 3$ and M is an oriented hypersurface in R^{n+1} with positive Riemannian curvature, then M is rigid.*

Proof. Let $f\colon M \to M'$ be an isometry and let $L' = L \circ f_*$. Since f is an isometry, the Gauss curvature equations give $R(X, Y)Y = R'(X, Y)Y$ or $\langle LY, Y\rangle L(X) - \langle LY, X\rangle L(Y) = \langle L'Y, Y\rangle L'(X) - \langle L'Y, X\rangle L'(Y)$, where X, Y in M_m. Choose an orthonormal base $X_1,\ldots,X_n$ of vectors at m, and let $LX_i = k_i X_i$. We show L' is invariant on each subspace P_{ij} spanned by X_i and X_j for $i \neq j$. Let $b_{rs} = \langle L'X_r, X_s\rangle$, and the Gauss curvature equations imply

$$k_i k_j X_i = b_{jj}L'X_i - b_{ij}L'X_j$$

$$k_i k_j X_j = -b_{ij}L'X_i + b_{ii}L'X_j.$$

Then $K(P_{ij}) = k_i k_j = b_{ii}b_{jj} - b_{ij}^2 > 0$ implies $L'X_i$ and $L'X_j$ lie in P_{ij}. Since $n \geq 3$, there is a third index r with $L'X_i$ in P_{ir}; hence $L'X_i$ lies in $P_{ir} \cap P_{ij}$, and thus X_i is an eigenvector of L'. For all i, let $L'X_i = h_i X_i$. Then $k_i k_j = h_i h_j > 0$ for all $i \neq j$; hence $k_i^2 = h_i^2$, so $h_i = \pm k_i$. The positive curvature condition also implies $h_i = k_i$ for all i, or $h_i = -k_i$ for all i. Thus $L = \pm L'$ and we apply the fundamental rigidity theorem.//

If in the above theorem we assume M is complete (or closed), then we need not assume it is oriented. For $n = 2$, the Cohn-Vossen theorem provides a similar result with the additional requirement that M be compact. We now examine some global functions and forms on an oriented surface M in R^3 before proving the Cohn-Vossen theorem.

Let N be the unit normal on M, let p in M, and let e_1, e_2 be a positively oriented orthonormal base field in the neighborhood U about p. Identifying p with the vector from the origin to p, define local functions y_1, y_2 on U, and a global function y_3 on M, by $p = y_1(p)e_1 + y_2(p)e_2 + y_3(p)N$. Define global 1-forms α and β on U by $\alpha(X) = \langle p, e_1\rangle\langle X, e_2\rangle - \langle p, e_2\rangle\langle X, e_1\rangle$ and $\beta(X) = \alpha(LX)$. One checks that α is independent of the particular positively oriented base e_1, e_2 used to define it, so α and β are global 1-forms on M. We now compute $d\alpha$ and $d\beta$. Let w_i, w_{ij} be the local forms belonging to the base e_1, e_2 so $w_{ij} = -w_{ji}$. Then $L(X) = \bar{D}_X(N) = w_{13}(X)e_1 + w_{23}(X)e_2$, $w_{i3} = b_{1i}w_1 + b_{2i}w_2$, and $b_{ij} = \langle Le_i, e_j\rangle$. Thus $\alpha = y_1w_2 - y_2w_1$ and $\beta = y_1w_{23} - y_2w_{13}$.

Since $y_i = \langle p, e_i\rangle$ we have $dy_i(X) = X\langle p, e_i\rangle = \langle \bar{D}_Xp, e_i\rangle + \langle p, \bar{D}_Xe_i\rangle = \langle X, e_i\rangle + \langle p, \Sigma_1^3 w_{ri}(X)e_r\rangle = w_i(X) + \Sigma_{r=1}^3 y_r w_{ri}(X)$. Thus, using the Cartan structural equations,

$$d\alpha = (w_1 + y_2w_{21} + y_3w_{31})w_2 - y_1(w_{21}w_1) - (w_2 + y_1w_{12} + y_3w_{32})w_1 +$$

$$+ y_2(w_{12}w_2) = 2w_1w_2 - y_3Hw_1w_2 = (2 - y_3H)v,$$

where v is the volume element. Similarly, $d\beta = (H - 2y_3K)v$.

If M is compact, then $\int_M(H - 2y_3K) = \int_M d\beta = \int_{\partial M}\beta = 0$, and $\int_M(2 - y_3H) = \int_M d\alpha = \int_{\partial M}\alpha = 0$, by Stokes' theorem. The equality $\int_M(H/2) = \int_M y_3K$ is called *Minkowski's formula*, and the other integral implies the area of M is $\int_M y_3(H/2)$. For other formulae of this type see Bormesen-Fenchel.

The above paragraph provides two examples of "Chern's formula for theorems in differential geometry," i.e., take a global 1-form w such that $dw = Fv$ where F is an "interesting" function, then state $\int_M F = 0$. Another example is that $\int_M K = 0$ is a necessary condition that w_{12} be a global 1-form.

THEOREM. (Cohn-Vossen). *A compact surface of positive Gaussian curvature is rigid.*

Proof. Let $f: M \to M'$ be an isometry of such surfaces, and assume the origin to be inside M so $y_3 > 0$. Let $L' = L_{M'} \circ f_*$, then L and L' are positive definite on M since $K = K' > 0$. Let $\Delta = \det(L - L')$. We show $L = \pm L'$ by showing $\Delta = 0$ and apply the following lemma: if A and B are two positive definite quadratic forms on R^2 with $\det A = \det B$, then $\det(A - B) \leq 0$, and $\det(A - B) = 0$ implies $A = \pm B$.

Let $\beta' = \alpha \flat L'$ and, as above, we compute $d\beta' = [H' - y_3(2K - \Delta)]v$. Hence $\int H' = \int y_3(2k - \Delta) = \int H - \int y_3\Delta$, all integrals taken over M. Thus $\int H' - \int H \geq 0$ since $y_3 \leq 0$, so $\int H' \geq \int H$. By symmetry we can reverse the inequality so $\int H' = \int H$ and $\int y_3 \Delta = 0$, which implies $\Delta = 0$.//

THEOREM. *If f is an isometry between two oriented surfaces that preserves the mean curvature and the third fundamental form, and the mean curvature is never identically zero on any neighborhood, then the surfaces are congruent.*

Proof. Let $f: M \to M'$ and let $L' = L_{M'} \circ f_*$. Equality of the third fundamental forms implies $\langle L^2X, Y\rangle = \langle (L')^2X, Y\rangle$ for all X, Y in M_m so $L^2 = (L')^2$. Using the characteristic equation for L and L' we have $HL = L^2 + KI = (L')^2 + K'I = H'L' = HL'$. Thus if $H(m) \neq 0$, then $L = L'$ at m, and since H never vanishes identically on any neighborhood, we have $L = L'$ on M by continuity.//

There is a theorem, similar to the preceding result, which states if f is a diffeo between two compact convex hypersurfaces that preserves the mean curvature and the third fundamental form, then the hypersurfaces are congruent or symmetric. For the proof of this result we refer the reader to Chern[1], p. 29. Problem 77 shows one can relax the compactness assumption in the Cohn-Vossen theorem by assuming the third fundamental form is preserved.

The above theorems were included chiefly for their accessibility. Much better theorems have been proved (see Pogorelov) with weaker differentiability assumptions.

Problems.

75. Prove that the volume V_n of the unit sphere in R^{n+1} is equal to $2^{n+1} \pi^k k!/n!$ for even $n = 2k$.

76. If M is a compact surface in R^3 with constant mean curvature and $y_3 > 0$ on M, show M is a sphere (see section 8.5).

77. If f is an isometry between two oriented surfaces in R^3 of positive Gaussian curvature which preserves the third fundamental form, show the surfaces are congruent or symmetric.

78. Use an integral argument to show there exists no compact minimal surface in R^3.

9. Existence Theory

Section 9.1. Involutive distributions and the Frobenius theorem.

We prove the standard theorem on the existence of "integral manifolds" of a distribution following Chevally (p. 88). The theorem also appears in Auslander and Mackenzie (p. 147) with the terminology altered slightly.

In this section let M be a C^∞ n-manifold. A *k-dimensional distribution* on a set A in M is a function P that assigns to each point p in A a k-dimensional subspace P_p of the tangent space M_p. We say P is C^∞ on A if A is open, and for each p in A there are k independent C^∞ vector fields $X_1,\dots,X_k$ which span P_m for all m in some neighborhood of p. A vector field X with domain B *lies in P* or *is in P* if $B \subset A$, and X_p is in P_p for all p in B. A C^∞ distribution P is *integrable* (*involutive* or *closed*) when it is closed under the bracket operation, i.e., if X and Y are any C^∞ fields with common domain that lie in P, then $[X, Y]$ lies in P. A submanifold V of M is an *integral submanifold* or *integral manifold* of P if V is contained in the domain of P, and $V_p = P_p$ for all p in V; thus the subspace of the tangent space M_p which belongs to V_p is exactly the subspace P_p.

The theorem proved below implies a C^∞ distribution has integral manifolds if and only if it is involutive. A slightly stronger statement is made involving the existence of a special coordinate system. First some terminology: if $x_1,\dots,x_n$ is a coordinate system on M with domain U, then define a *slice of U* to be any subset of U on which r of the functions $x_1,\dots,x_n$ are constant, where $1 \leqslant r < n$. Obviously, each slice of U is a submanifold of U (or M).

THEOREM. *Let P be a k-dimensional involutive C^∞ distribution on M. For any m in M there exists a coordinate system $x_1,\dots,x_n$ with domain U including m such that the coordinate fields $\partial/\partial x_j$ for $j = 1,\dots,k$ span P at each point of U. Thus the slices of U of which $x_{k+1},\dots,x_n$ are constant are integral manifolds of P.*

The theorem is proved by induction on k. The case $k = 1$ is covered by the following lemma, and note in this case any distribution is automatically involutive.

LEMMA. *If X be a C^∞ vector field on M, p in M, and $X_p \neq 0$, then there exists a coordinate system $y_1,\dots,y_n$ on a neighborhood U of p with $X = \partial/\partial y_1$ on U.*

Proof (of lemma). Let $x_i = u_i \circ \phi$ be a coordinate system on the neighborhood V of p with $x_i(p) = 0$ and $\partial/\partial x_1(p) = X_p$. Let $X = \Sigma_1^n a_i(\partial/\partial x_i)$ where a_i are C^∞ real valued functions on V and $a_1(p) \neq 0$, and restrict V if necessary so $a_1 \neq 0$ on V. Setting up the system of differential equations for the integral curves σ of X on V, we have

$$\frac{d(x_i \circ \sigma)}{dt} = a_i \circ \sigma \qquad \text{or} \qquad \frac{df_i}{dt} = a_i(f_1(t),\dots,f_n(t))$$

where $f_i(t) = x_i \circ \sigma(t)$. Applying an existence theorem from the theory of differential equations (Coddington and Levinson, Chapter 1) we obtain an $r > 0$ and n functions $F_i(t, a_1, a_2,\dots,a_n)$ which are C^∞ on the neighborhood W of the origin in R^{n+1} where $|t| < r$ and $|a_i| < r$ such that for $i = 1,\dots,n$:

(1) $F_i(0, a_1, a_2,\dots,a_n) = a_i$,

(2) $(F_1(b),\dots,F_n(b))$ in $\phi(V)$ for b in W,

(3) Letting $F(t, a_2,\dots,a_n) = \phi^{-1}[F_1(t, 0, a_2,\dots,a_n),\dots,F_n(t, 0, a_2,\dots,a_n$ define a map F of $B(0, r)$ in R^n into V; then for fixed $a_2,\dots,a_n$ the curves

$$\sigma_{(a_2,\dots,a_n)}(t) = F(t, a_2,\dots,a_n)$$

are integral curves of X, i.e., $F_*(\partial/\partial u_1) = X$.

For points $(0, a_2,\ldots,a_n)$ in $B(0, r)$ we notice that $F(0, a_2,\ldots,a_n) = \phi^{-1}(0, a_2,\ldots,a_n)$; hence $F_*(\partial/\partial u_i)_{\text{origin}} = \partial/\partial x_i(p)$ for $i = 2,\ldots,n$. Since $F_*(\partial/\partial u_1)_b = X_{F(b)}$ for all b in $B(0, r)$ we have $F_* = (\phi^{-1})_*$ at the origin in R^n. Hence F_* is non-singular at the origin and by the Inverse Function Theorem F is a diffeo between a neighborhood of the origin and a neighborhood U of p with $U \subset V$. Finally, let $y_i = u_i \circ F^{-1}$ on U.//

Intuitively, in the above proof we have changed the $x_1,\ldots,x_n$ coordinates about p by leaving the slice where $x_1 = 0$ fixed, and replacing the "x_1-coordinate curves" by integral curves of X emanating from this slice.

Proof (of theorem). Take the point m and take C^∞ fields $X_1,\ldots,X_k$ that span P on a neighborhood U_1 on m. Apply the previous lemma to get a coordinate system $y_1,\ldots,y_n$ about m with domain $U_2 \subset U_1$ such that $\partial/\partial y_1 = X_1$ on U_2, and assume $y_i(m) = 0$.

If $k = 1$, then the coordinate system $y_1,\ldots,y_n$ satisfies the conclusion of the theorem. If $k > 1$, we assume the theorem is true for distributions of dimension less than k, and we define the $(k - 1)$ dimensional distribution $\overline{P}$ on U_2 by $\overline{P}_p = [X \text{ in } P_p\colon\ X_p y_1 = 0]$ for p in U_2. This is a $(k - 1)$-dimensional C^∞ distribution for it is spanned by the $(k - 1)$ independent C^∞ fields $Y_i = X_i - (X_i y_1)X_1$ for $i = 2,\ldots,k$. It is involutive since if Y and Z are in $\overline{P}$, then $[Y, Z]$ is in P and $[Y, Z]y_1 = Y(Zy_1) - Z(Yy_1) = 0$ on U_2, so $[Y, Z]$ is in $\overline{P}$.

Let V_0 be the slice of U_2 defined by $y_1 = 0$. Then for p in V_0, $\overline{P}_p \subset (V_0)_p$, so we apply the induction hypothesis to the distribution $\overline{P}$ on the manifold V_0 to obtain a coordinate system $z_2,\ldots,z_n$ on the neighborhood U_3 about m in V_0 such that $\partial/\partial z_2,\ldots,\partial/\partial z_k$ span $\overline{P}$ on U_3. We define the map $\pi\colon U_2 \to V_0$ by $\pi(p) = \phi^{-1}(0, y_2(p),\ldots,y_n(p))$, where ϕ is the coordinate map so $y_i = u_i \circ \phi$. Let $U_4 = \pi^{-1}(U_3)$ and define functions $x_1,\ldots,x_n$ on U_4 by $x_1 = y_1$, $x_2 = z_2 \circ \pi,\ldots,x_n = z_n \circ \pi$. Then the functions $x_1, x_2,\ldots,x_n$ define a coordinate system in a neighborhood U of m with $U \subset U_4$; indeed, $\partial/\partial x_1(m) = \partial/\partial y_1(m)$, while $\partial/\partial x_2,\ldots,\partial/\partial x_n$ span $(V_0)_m$ at m.

We show $\partial/\partial x_1,\ldots,\partial/\partial x_k$ span P on U by showing they span the same subspace as $X_1, Y_2,\ldots,Y_k$. Let $Y_1 = X_1$, then we show $Y_i x_j = 0$ for $i = 1,\ldots,k$ and $j = k + 1,\ldots n$. Since $Y_1 = X_1 = \partial/\partial x_i$, we immediately see $Y_1 x_j = 0$ for $j \neq 1$. Since P is involutive, there are C^∞ functions g_{irs} on U such that for $i \leq k$ and $r \leq k$ we have $[Y_i, Y_r] = \sum_{s=1}^{k} g_{irs} Y_s$. Thus for $i = 2,\ldots,k$ and $j > k$, $Y_1(Y_i x_j) = [Y_1, Y_i]x_j = \sum_{s=1}^{k} g_{1is}(Y_s x_j)$. This implies the functions $Y_i x_j$ satisfy a linear homogeneous system of ordinary differential equations along any x_1-curve. But on V_0, $x_j = z_j$ for $j > 1$ and $Y_i x_j = Y_i z_j = 0$ on V_0 for $j > k$ because of the choice of the coordinates $z_2,\ldots,z_n$. Hence, by the uniqueness of solutions to systems of the above type, $Y_i x_j = 0$ for $i \leq k$ and $j > k$.//

We use the theorem on involutive distributions to prove the classical Frobenius theorem on (total) partial differential equations (see Levi-Civitae). This theorem can be stated roughly as follows: there exist unique solution functions $f_i(x_1,\ldots,x_k)$, with prescribed values at a point, to the system of partial differential equations

$$\frac{\partial f_i}{\partial x_j} = A_{ij}(x_1,\ldots,x_k, f_1,\ldots,f_d)$$

if and only if for all $j \leq k$, $r \leq k$ and $i \leq d$,

$$\frac{\partial A_{ij}}{\partial x_r} + \sum_{s=1}^{d} \frac{\partial A_{ij}}{\partial f_s} A_{sr} = \frac{\partial A_{ir}}{\partial x_j} + \sum_{s=1}^{d} \frac{\partial A_{ir}}{\partial f_s} A_{sj}$$

(which is merely what the chain rule demands if $\dfrac{\partial^2 f_i}{\partial x_r \partial x_j} = \dfrac{\partial^2 f_i}{\partial x_j \partial x_r}$).

THEOREM (Frobenius). *For $1 \leq i \leq d$ and $1 \leq j \leq k$, let $A_{ij}(x_1,\ldots,x_k, u_1,\ldots,u_d)$ be C^∞ real valued functions on an open set Q in R^n where $n = k + d$, and we have labelled the coordinate functions of R^n in order to conveniently express partial derivatives. Let $(a; b) = (a_1,\ldots,a_k, b_1,\ldots,b_d)$ be in Q. Then there exists a unique set of C^∞ real valued functions $f_1,\ldots,f_d$ defined on a neighborhood V of a and satisfying the following three conditions:*

(1) *$f_i(a) = b_i$ or $f(a) = b$, where f is the mapping of V into R^d defined by $f(p) = (f_1(p),\ldots,f_d(p))$,*

(2) *if p in V, then $(p: f(p))$ in Q, and*

(3) *if p in V, then* $\frac{\partial f_i}{\partial x_j}(p) = A_{ij}(p; f(p))$,

iff at every point of Q,

$$(4) \quad \frac{\partial A_{ij}}{\partial x_r} + \Sigma_{s=1}^{d} \frac{\partial A_{ij}}{\partial u_s} A_{sr} = \frac{\partial A_{ir}}{\partial x_j} + \Sigma_{s=1}^{d} \frac{\partial A_{ir}}{\partial u_s} A_{sj}.$$

Proof. Let $e_1,\ldots,e_n$ be the usual global orthonormal vector fields on R^n. We use the functions A_{ij} to define C^∞ vector fields $Y_1,\ldots,Y_k$ on Q by $Y_r = e_r + \Sigma_{s=1}^{d} A_{sr}e_{k+s}$. These vector fields are independent at each point of Q and hence they span a k-dimensional C^∞ distribution P on Q. We form brackets,

$$[Y_r, Y_q] = [e_r + \Sigma_{s=1}^{d} A_{sr}e_{k+s}, e_q + \Sigma_{t=1}^{d} A_{tq}e_{k+t}]$$

$$= \Sigma_{t=1}^{d}\left(\frac{\partial A_{tq}}{\partial x_r} + \Sigma_{s=1}^{d} A_{sr}\frac{\partial A_{tq}}{\partial u_s}\right)e_{k+t} - \Sigma_{s=1}^{d}\left(\frac{\partial A_{sr}}{\partial x_q} + \Sigma_{t=1}^{d} A_{tq}\frac{\partial A_{sr}}{\partial u_t}\right)e_{k+s},$$

and thus by condition (4), $[Y_r, Y_q] = 0$.

Hence the distribution P is involutive and by the theorem above there exists an integral manifold U of T through $(a; b)$ with $U \subset Q$. Let ϕ: $U \to R^k$ by $\phi(a'; b') = a'$, then $\phi_*(Y_r) = e_r$ and ϕ_* is nonsingular on the tangent space of U at $(a; b)$. Thus there is a neighborhood V of a and a map F which is a diffeo of V onto $F(V) \subset U$ such that $F \circ \phi$ and $\phi \circ F$ give the identity map on $F(V)$ and V, respectively. Define $f_1,\ldots,f_d$ on V by $F(p) = (p; f_1(p),\ldots,f_d(p))$. Then the functions $f_1,\ldots,f_d$ are C^∞ functions satisfying (1) and (2), and (3) follows since $F_*(e_r) = Y_r$ for $r \leq k$.

The implication of the theorem in the other direction is trivial.//

Actually the Frobenius theorem in turn can be used to prove the theorem on involutive distributions. A k-dimensional distribution P about m can be carried to an open set Q in R^n via a coordinate map. Furthermore one may choose the coordinate map so the induced distribution on Q is spanned by vectors $Y_1,\ldots,Y_k$ of the type defined above, and this defines functions A_{ij}. The involutive condition will then imply $[Y_r, Y_q] = 0$, since $[Y_r, Y_q]$ must be a linear combination of $Y_1,\ldots,Y_k$ at each point. This implies the integrability condition (4) of the Frobenius theorem is satisfied which we then apply to get local integral manifolds. One actually has to state the Frobenius theorem to include the C^∞ dependence of the solution functions on the initial conditions (which follows from the Chevalley theorem) in order to obtain the full equivalence.

A first application of the Frobenius theorem provides a useful theorem concerning the existence of coordinate systems.

THEOREM. *Let M be an n-dimensional C^∞ manifold and let $X_1,\ldots,X_n$ be a set of independent C^∞ vector fields on a neighborhood U of m in M. Then there exists a set of coordinate functions $x_1,\ldots,x_n$ defined on a neighborhood V of m with $V \subset U$ and $X_i = \partial/\partial x_i$ on V for all i iff $[X_i, X_j] = 0$ for all i and j.*

Section 9.2. *The fundamental existence theorem for hypersurfaces.*

Let U be an open set in R^n on which is defined real valued C^∞ functions g_{ij} and b_{ij} for $1 \leq i, j \leq n$ such that the matrices (g_{ij}) and (b_{ij}) are symmetric and (g_{ij}) is positive definite. Roughly speaking, we prescribe conditions which imply the existence of a coordinate system on a hypersurface of R^{n+1} such that the matrices (g_{ij}) and (b_{ij}) are the coordinate representations of the first and second fundamental forms, respectively. We demand that (g_{ij}) and (b_{ij}) satisfy the Gauss curvature and Codazzi-Mainardi equations, and explain this demand. On U, define functions Γ^i_{jk} in terms of the g_{ij} by the classica formula (see section 6.2), and define functions $w_{ij}(e_k) = \Gamma^i_{jk}$, $w_{n+1j}(e_k) = -b_{jk}$, $w_{jn+1}(e_k) = \Sigma_{r=1}^n (g^{-1})_{jr} b_{rk}$, and $w_{n+1n+1}(e_k) = 0$, for all $i, j, k \leq n$. Then if there was a coordinate system with coordinate fields $e_1,\ldots e_n$ whose image set was U, the Gauss curvature equations and Codazzi-Mainardi equations imply (see section 6.6)

$$(1) \qquad dw_{ij}(e_r, e_s) = -\Sigma_{k=1}^{n+1} w_{ik} \wedge w_{ks}(e_r, e_s)$$

and

$$(2) \qquad dw_{jn+1}(e_r, e_s) = -\Sigma_{k=1}^{n} w_{jk} \wedge w_{kn+1}(e_r, e_s),$$

respectively. Thus we say (g_{ij}) and (b_{ij}) *satisfy the Gauss curvature and Codazzi-Mainardi equations* if (1) and (2) hold for the functions defined on U where the left sides are computed by $dw_{ij}(e_r, e_s) = (\partial/\partial u_r)w_{ij}(e_s) - (\partial/\partial u_s)w_{ij}(e_r)$, etc.

THEOREM. *Let (g_{ij}) and (b_{ij}) be defined on U as described above and suppose they satisfy the Gauss curvature and Codazzi-Mainardi equations. Then for any point p in U, there is a neighborhood $V \subset U$ and a C^∞ mapping F: $V \to R^{n+1}$ such that $F(V)$ is an n-dimensional submanifold of R^{n+1}, F^{-1} is a coordinate map on $F(V)$, and (g_{ij}) and (b_{ij}) are the coordinate representation matrices of the first and second forms on $F(V)$, respectively.*

Proof. Let $u_1,\ldots,u_n$ be the natural coordinate functions on U. We seek $(n+1)R^{n+1}$-valued functions $e_1,\ldots,e_{n+1}$ defined on U that satisfy the Gauss equations and Weingarten equations, i.e.,

$$\text{(3)} \qquad \frac{\partial e_i}{\partial u_j} = \Sigma_{k=1}^{n+1} w_{ki}(e_j)e_k = (\bar{D}_{e_j}(e_i))$$

where $j = 1,\ldots,n$ and $i = 1,\ldots,n+1$. Each of the equations in (3) has $n+1$ components, and the differentiation operator $\partial/\partial u_j$ is applied to each component. In order to apply the Frobenius' theorem we compute $\partial^2 e_i/\partial u_k \partial u_j$, using (3) to obtain

$$\frac{\partial^2 e_i}{\partial u_k \partial u_j} = \Sigma_r\left(\frac{\partial w_{ri}(e_j)}{\partial u_k} e_r + w_{ri}(e_j)\frac{\partial e_r}{\partial u_k}\right)$$

$$= \Sigma_r\left(\frac{\partial w_{ri}(e_j)}{\partial u_k} e_r + \Sigma_s w_{ri}(e_j) w_{sr}(e_k) e_s\right)$$

where we sum r and s from 1 to $n+1$. The integrability conditions are the equations

$$\text{(4)} \quad \frac{\partial w_{ri}(e_j)}{\partial u_k} + \Sigma_s w_{si}(e_j)w_{rs}(e_k) = \frac{\partial w_{ri}(e_k)}{\partial u_j} + \Sigma_s w_{si}(e_k)w_{rs}(e_j),$$

which follows from (1) and (2).

At the origin in R^{n+1} we choose initial vectors $e_1,\dots,e_{n+1}$, so $\langle e_i, e_j\rangle = g_{ij}(p)$ for $i, j \leq n$, $\langle e_i, e_{n+1}\rangle = g_{in+1}(p) = 0$, $e_{n+1} = \partial/\partial u_{n+1}$, and $e_1,\dots,e_{n+1}$ positively oriented. Applying the Frobenius theorem we obtain a neighborhood V_1 of p and $(n+1)R^{n+1}$-valued C^∞ functions $e_1,\dots,e_{n+1}$ that satisfy (3).

To check that $\langle e_i, e_j\rangle = g_{ij}$ and $\langle e_i, e_{n+1}\rangle = 0$ at all points on V_1 for $i, j \leq n$, we must again apply the Frobenius theorem. Let $G_{ij} = \langle e_i, e_j\rangle$ on V_1 for $i, j \leq n+1$. Then by (3) and the product rule we have

$$(5) \qquad \frac{\partial G_{ij}}{\partial u_k} = \Sigma_{r=1}^{n+1}[w_{ri}(e_k)G_{rj} + w_{rj}(e_k)G_{ir}]$$

on V_1. But from the definition of $w_{ri}(e_k) = \Gamma^r_{ik}$ in terms of g_{ij} we find the functions g_{ij} also satisfy (5) where we define $g_{in+1} \equiv \delta_{in+1}$. By using (1) and (2) we verify the system (5) satisfies the necessary integrability conditions for the Frobenius theorem and since $G_{ij}(p) = g_{ij}(p)$ we have $G_{ij} = g_{ij}$ on a neighborhood V_2 on p.

Define functions A_{ij} on V_2 for $i = 1,\dots,n+1$ and $j = 1,\dots,n$ by $e_j = (A_{ij},\dots,A_{n+1j})$, and consider the system of equations

$$(6) \qquad \frac{\partial f_i}{\partial u_j} = A_{ij}.$$

Here $\dfrac{\partial A_{ij}}{\partial u_k} = \dfrac{\partial A_{ik}}{\partial u_j}$ since $\dfrac{\partial e_j}{\partial u_k} = \dfrac{\partial e_k}{\partial u_j}$ (for $\Gamma^i_{jk} = \Gamma^i_{kj}$). Thus, letting $f_i(p) = 0$ for $i = 1,\dots,n+1$, we apply the Frobenius theorem again to get C^∞ functions $f_1,\dots,f_{n+1}$ on a neighborhood V_3 of p with $V_3 \subset V_2$.

Finally, we define $F\colon V_3 \to R^{n+1}$ by $F(m) = (f_1(m),\dots,f_{n+1}(m))$ for m in V_3. Then F is C^∞ and $F_*(\partial/\partial u_j(m)) = e_j(m)$ for $j = 1,\dots,n$. Thus F is a diffeo of a neighborhood V of p onto its image $F(V)$ in R^{n+1} and $V \subset V_3$. The map F^{-1} is a coordinate system on $F(V)$ with coordinate vectors $e_1,\dots,e_n$, so $g_{ij} = \langle e_i, e_j\rangle$ and $\langle Le_i, e_j\rangle = \langle \Sigma_r w_{rn+1}(e_i)e_r, e_j\rangle = \Sigma_r w_{rn+1}(e_i)g_{rj} = \Sigma_{r,s}(g^{-1})_{rs}b_{si}g_{rj} = b_{ji}$ as desired.//

Section 9.3. The exponential map.

Let D be a connexion on M^n. From section 5.1 we know for each vector X, tangent to M at m, there is unique geodesic $g_X(t)$ of the connexion D, which is defined on a neighborhood of zero in R with $g_X(0) = m$ and tangent X at $t = 0$. Furthermore, for appropriate s in R, $g_{sX}(t) = g_X(st)$ by the nature of the differential equations defining the geodesics. This implies that $g_{aX}(1)$ is defined if $g_X(a)$ is defined, thus $g_Y(1)$ is a well-defined point of M for Y near zero in M_m.

Definition. *For Y in M_m, we define $\exp_m Y = g_Y(1)$ when the latter is defined. The map $\exp_m$ is called the* exponential map.

The name "exponential map" is used because in a special case for the general linear group $GL(n, R)$ it becomes the classical map, $A \to e^A = I + A + (A^2/2!) + \ldots$, from the set of all n by n real matrices into the set of non-singular matrices (problem 81).

Our current objective is to obtain some important properties of the exponential map and to state these precisely we must use the tangent bundle $T(M)$ of M.

Proposition 1. Let N be the subset of $T(M)$ such that if (m, Y) in N then $\exp_m(Y)$ is defined and define the map exp: $N \to M$ by $\exp(m, Y) = \exp_m(Y)$. Then N is an open set and exp is C^∞ on N. In particular let $\hat{M} = [(m, 0)$ in $T(M)$: m in $M]$, then there is an open set $\hat{N}$ in $T(M)$ such that $\hat{M} \subset \hat{N} \subset N$.

Proof. We do not completely prove the above proposition. Applying the local theory of differential equations we prove the last statement of the theorem and we prove exp: $\hat{N} \to M$ is C^∞. Then we sketch the proof that exp is C^∞ on N and refer the reader to Lang.

Using the above notation, if $g(t)$ is a geodesic in the neighborhood U, then

$$\frac{d^2(x_k \circ g)}{dt^2} + \Sigma_{i,j=1}^n \Gamma_{ij}^k \frac{d(x_i \circ g)}{dt} \quad \frac{d(x_j \circ g)}{dt} = 0$$

for t such that $g(t)$ in U. For each point $(m, 0)$ in $T(M)$ take a coordinate neighborhood U_m of m in M and apply the existence and uniqueness theorem to the above differential equation to obtain a real number $b > 0$, a neighborhood V_m of $(m, 0)$ in $T(M)$ with $V_m \subset \pi^{-1}(U_m)$, a C^∞

map g: $(-b, b) \times V_m \to M$, such that for fixed (p, Y) in V_m, the curve $g_Y(t) = g(t; p, Y)$ is the unique geodesic defined on $(-b, b)$ which passes through p with tangent Y at $t = 0$. Moreover, for (p, Y) in V_m and $a > 0$, we have g_{aY} defined on $(-b/a, b/a)$, since $g_{aY}(t) = g_Y(at) = g(at; p, Y)$. Choose $a > 0$ so $a < b$ and let $W_m = [(p, X)$ in V_m: $(p, X/a)$ in $V_m]$. Then for (p, X) in W_m, $\exp(p, X) = g(1; p, X) = g(a; p, X/a)$ is defined and exp is C^∞ on W_m.

Let $\hat{N} = U_m W_m$, and the last sentence of the theorem is proved.

For each (p, Y) in $T(M)$, choose $a > 0$ so (p, aY) in $\hat{N}$, and thus $g_Y(t)$ is defined in some neighborhood of $t = 0$. As usual, for any curve g let T_g be its tangent vector and define *the natural lifting of a curve g in M to a curve $\bar{g}$ in* $T(M)$ by $\bar{g}(t) = (g(t), T_g(t))$. Now define a *global vector field Z* on $T(M)$ by $Z(p, Y) = T_{\bar{g}_Y}(0)$. Then Z is a C^∞ field on $T(M)$ by the above analysis, and if σ is an integral curve of Z, then $\pi \circ \sigma$ is a geodesic in M. The field Z is called the *geodesic flow field* associated with the connexion. The fact that exp is C^∞ on all of N now follows from Theorem 5 on p. 66 in Lang.//

Corollary 1. For fixed p in M, the map $\exp_p$ is a diffeo of a neighborhood of 0 in M_p onto a neighborhood of p. Furthermore, if η_0: $M_p \to (M_p)_0$ is the natural map of the tangent space at p onto *its* tangent space at 0, then $(\exp_p)_* \circ \eta_0$ is the identity map on M_p.

Proof. The map η_0 is defined by choosing any base $e_1,\dots,e_n$ of M_p and letting $z_1,\dots,z_n$ be its dual base. Then $z_1,\dots,z_n$ are a global coordinate system on the vector space viewed as a C^∞ manifold. Let $\eta_0(e_i) = (\partial/\partial z_i)_0$ for all i. This map η is independent of the particular base $e_1,\dots,e_n$; furthermore, by evaluating the global fields $\partial/\partial z_i$ at any point Y in M_m, we obtain a natural isomorphism η_Y: $M_p \to (M_p)_Y$ In these notes, for any X in M_p we let $\bar{X}$ be the *natural constant vector field on M_p* associated with X, where $\bar{X}_Y = \eta_Y(X)$.

Take X in M_p, then $\eta_0(X) = \bar{X}_0$. To compute $(\exp_p)_*\bar{X}_0$ we note $\bar{X}_0$ is the tangent vector at $t = 0$ to the ray $\gamma(t) = tX$ in M_p. The curve $\exp_p \circ \gamma(t) = \exp_p tX$ is by definition the geodesic through p with tangent X at $t = 0$. Thus $(\exp_p)_*\bar{X}_0 = X$. Thus $(\exp_p)_*$ is non-singular and onto at the origin in M_p. The corollary now follows by applying the Inverse Function theorem.//

Corollary 2. Let G: $\hat{N} \to M \times M$ by $G(p, Y) = (p, \exp_p Y)$. Then G is C^∞ and G_* is non-singular and onto at all points $(p, 0)$ in $T(M)$.

Proof. Let π_i: $M \times M \to M$ by $\pi_i(m_1, m_2) = m_i$ for $i = 1, 2$. Each π_i is C^∞. Since $\pi_1 \circ G = \pi_1$ and $\pi_2 \circ G = \exp$, the map G is C^∞ on $\hat{N}$.

The tangent space $\hat{N}_{(p,0)}$ is naturally isomorphic to $M_p \times (M_p)_0$, while the tangent space to $M \times M$ at $G(p, 0) = (p, p)$ is naturally isomorphic to $M_p \times M_p$. In terms of these natural isomorphic spaces, G_* is the identity on the first factor and $(\exp_p)_*$ on the second factor. Hence, by Corollary 1, G_* is non-singular at $(p, 0)$.//

We apply Corollary 1 to obtain *normal coordinate systems.* For any m in M let $e_1,\ldots,e_n$ be a base of M_m, let $z_1,\ldots,z_n$ be its dual base, and let $\bar{U}$ and U be neighborhoods of O in M_m and m in M, respectively, such that $\exp_m$ is a diffeo of $\bar{U}$ onto U whose inverse we denote by $\exp^{-1}$. Then define C^∞ functions $x_1,\ldots,x_n$ on U by $x_i = z_i \circ \exp^{-1}$ for all i. These functions $x_1,\ldots,x_n$ define a *normal coordinate system* (of the connexion D) on U. The curves σ in U such that $x_i \circ \sigma(t) = a_i t$ for constants $a_1,\ldots,a_n$, are geodesics emanating from m(at $t = 0$), and if Γ^i_{jk} are the connexion functions on U for this coordinate system, then $\Gamma^i_{jk}(m) = 0$, provided the connexion has zero torsion.

One verifies this last statement by letting $X_i = \partial/\partial x_i$ and then $D_{X_i}(X_j) = \Sigma^n_{k=1}\Gamma^k_{ji}X_k$ by definition of Γ^k_{ji}. Since the curve σ with $x_i \circ \sigma(t) = t$, $x_j \circ \sigma(t) = t$, and $x_k \circ \sigma(t) = 0$ for $k \neq i$ or j, is a geodesic, its tangent $X_i + X_j$ satisfies the condition $0 = D_{(X_i+X_j)}(X_i + X_j) = D_{X_i}X_i + D_{X_i}X_j + D_{X_j}X_i + D_{X_j}X_j$. Thus at m, $D_{X_i}X_i = 0$ for all i, since each coordinate curve emanating from m is a geodesic, and if D has zero torsion, then $0 = 2(D_{X_i}X_j)_m = 2\Sigma_k\Gamma^k_{ji}(m)X_k$ so $\Gamma^k_{ji}(m) = 0$ for all i, j, and k.

We apply Corollary 2 to obtain *Fermi coordinates along a curve.* Let σ be a C^∞ curve in M that is univalent on the open interval $I \subset R$. Let $e_1,\ldots,e_n$ be C^∞ fields on σ that are independent at each $\sigma(t)$ and $e_n(t) = T_\sigma(t)$ for all t in I. Let $z_1,\ldots,z_n$ be the dual base to $e_1,\ldots,e_n$ for each t. By Corollary 2, there is a neighborhood V of $\hat{M} \subset T(M)$ such that G is a diffeo of V onto a neighborhood N_M of the diagonal in $M \times M$. Let $U = [(m, Y)$ in V: $m = \sigma(t)$ and $z_n(Y) = 0$ for some t in I]. Then $F = G|_U$ is a 1 to 1 C^∞ map of the submanifold U into $M \times M$. Moreover F_* is non-singular at each point of U, so F is an imbedding of U into $M \times M$. The map $H = \pi_2 \circ F$ then gives a 1 to 1 C^∞ map of U onto an open neighborhood W of the image set $\sigma(I)$. Define Fermi coordinate $y_1,\ldots,y_n$ on p in W by letting $H^{-1}(p) = (\sigma(t), Y)$ in W and $y_i(p) = z_i(Y)$ for $i = 1,\ldots,n-1$ and $y_n(p) = t$.

More special types of Fermi coordinates can be defined by taking $e_1,\ldots,e_n$ to be a parallel base along a geodesic σ, and in the Riemannian case, one can take an orthonormal parallel base along a geodesic.

Section 9.4. Convex neighborhoods.

This section is devoted to proving the following theorem, due to J. H. C. Whitehead.

THEOREM. *Let M be a C^∞ manifold and D be a C^∞ connexion on M. Then for any point m in M there is a neighborhood U of m which is convex; i.e., for any two points in U there is a unique geodesic of D which joins the two points and lies in U.*

Proof. We may assume D has zero torsion, since by section 5.4 there is a unique torsion-free connexion with the same geodesics. The theorem is local, and we work completely in one coordinate neighborhood of m. From the previous section we choose a normal coordinate system $x_1,\ldots,x_n$ about m with domain A, thus $x_i(m) = 0$ and $\Gamma^i_{jk}(m) = 0$ for all i, j, and k. Let $d(p, q)$ be a local metric on A defined by $d(p, q) = [\Sigma_i(x_i(p) = x_i(q))^2]^{1/2}$, $f(p) = d(p, m)$, and let $B(p, c) = [q$ in A: $d(p, q) < c]$ for p in A. Also for p in A and X in M_p, let $|X| = [\Sigma_i dx_i(X)^2]^{1/2}$.

By Corollary 2 in section 3, for each p in A there is a real number $r_p > 0$ so that G is a diffeo on the set (q, X) where q in $B(p, r_p)$ and $d(q, \exp_p X) < r_p$. Take $c > 0$ so $\bar{B} = \bar{B}(m, c) \subset A$. For each p in $\bar{B}$ we obtain an integer $r_p > 0$. The family of neighborhoods $B(p, r_p)$ for p in $\bar{B}$ is a covering of the compact set $\bar{B}$, hence we may select a finite subcovering of neighborhoods belonging to $p_1,\ldots,p_k$. Let $s = \min[r_1,\ldots,r_k]$. Then for any p in $\bar{B}$, $\exp_p$ maps a neighborhood $\bar{U}_p$ of the origin in M_p diffeo onto $B(p, s)$. This follows since p in $B(p_j, r_j)$ for some j and hence G is a diffeo on the set (q, X) for q in $B(p_j, r_j)$ and $d(q, \exp_p X) < r_j$. We fix $q = p$, and $\exp_p$ is a diffeo of a neighborhood $\bar{V}_p$ of 0 in M_p onto $B(p, r_j)$ and $s \leq r_j$. We have proved the following:

LEMMA 1. *For any* $c > 0$ *with* $\bar{B}(m, c) \subset A$, *there exists an* $s > 0$ *such that for* p *in* $\bar{B}(m, c)$ *the map* $\exp_p$ *is a diffeo from a neighborhood* $\bar{U}_p$ *of* 0 *in* M_p *onto* $B(p, s) \subset A$.

We now prove two lemmas that complete the proof of the theorem.

LEMMA 2. *There exists a real number* a, $0 < a < 1$ *and* $\bar{B}(m, a) \subset A$, *such that if* $0 < b < a$ *and* g *is a geodesic with tangent* T_g *and* $f \circ g(0) = b$, $T_{g(0)}f = 0$, *then* $f \circ g$ *has a strict relative minimum at* $g(0)$. *Thus if* g *is tangent to the "sphere about* m *of radius* b" *at* $g(0)$, *then* g *lies outside of* $B(m, b)$ *near* $g(0)$.

Proof. We may assume $|T_{g(0)}| = 1$. Let $T = \Sigma_j a_j X_j$ where $X_j = \partial/\partial x_j$ and $a_j \circ g = (d/dt)(x_j \circ g)$, and we assume T is extended to a C^∞ field in a neighborhood of $g(0)$. Since $f = [\Sigma_i x_i^2]^{1/2}$ we have $Tf = \Sigma_j a_j (X_j f) = (1/f)\Sigma_j a_j x_j$ and $T^2 f = \Sigma_k a_k [(-x_k/f^3)(\Sigma_j a_j x_j) + (1/f)(\Sigma_j (X_k a_j) x_j + a_k)]$. At $t = 0$, or at $g(0)$, $Tf = 0$; hence $T^2 f = (1/b)[\Sigma_k a_k^2 + \Sigma_{k,j} a_k x_j (X_k a_j)]$. But at $g(0)$, $\Sigma_k a_k^2 = |T|^2 = 1$ and $\Sigma_k a_k (X_k a_j) + \Sigma_{r,s} \Gamma^j_{rs} a_r a_s = 0$, since g is a geodesic. Thus $T^2 f = (1/b)[1 - \Sigma_{j,r,s} x_j \Gamma^j_{rs} a_r a_s]$. Choose $a > 0$ and $a < 1$ so for points p with $f(p) \leq a$, $|\Gamma^i_{jk}(p)| < 1/2n^3$ for all i, j, and k, which is possible since Γ^i_{jk} continuous and $\Gamma^i_{jk}(m) = 0$. Then at $g(0)$,

$$|\Sigma_{j,r,s} x_j \Gamma^j_{rs} a_r a_s| \leq (1/2n^3)(\Sigma_{j,r,s,} 1) \leq 1/2,$$

hence $T^2 f(g(0)) > 0$, which implies $f \circ g$ has a strict relative minimum at 0.//

LEMMA 3. *Let* a *be given by Lemma 2 and apply Lemma 1 with* $c = a/2$ *to obtain* $s > 0$ *with* $s < (2/3)a$. *Then* $B(m, s/2)$ *is convex.*

Proof. Choose any p and q in $B(m, s/2)$. By lemma 1 there is a geodesic g defined on some interval $[0, u]$ with $g(0) = p$, $g(u) = q$, and $g(t)$ in $B(p, s)$ for all t in $[0, u]$. We show $f \circ g(t) < s/2$ for all t in $[0, u]$. Let v be a number in $[0, u]$ where $f \circ g$ attains its maximum value. Then $f \circ g(v) < a$ since $f \circ g(v) = d(m, g(v)) \leq d(m, p) + d(p, g(v)) < (s/2) + s < a$. Suppose $f \circ g(v) \geq s/2$. Then $(f \circ g)'(v) = 0$ and $f \circ g(v) < a$ which implies by Lemma 2 that $f \circ g$ has a strict relative minimum at v which gives a contradiction. Thus $B(m, s/2)$ is convex.//

Section 9.5. Special coordinate systems.

Let M be a Riemannian n-manifold, let ϕ be a coordinate map on M with domain U and $x_i = u_i \circ \phi$, and let $X_i = \partial/\partial x_i$. The coordinate system $x_1,\ldots,x_n$ is *orthogonal* if $\langle X_i, X_j\rangle = 0$ for $i \neq j$. If the map ϕ is a conformal map of U into R^n (with respect to the canonical Riemannian metric on R^n) then the coordinate system is *isothermal* or *conformal* (and hence also orthogonal). When M^n is a hypersurface in some $\bar{M}^{n+1}$, the coordinate system is *principal* if each X_i is a principal vector, and it is *asymptotic* if each X_i is an asymptotic vector.

In this section we study the existence of such special coordinate systems when $n = 2$. Orthogonal systems and conformal systems exist about any point, and the latter may be used to define a Riemann surface structure on M. Principal coordinates exist of necessity about any non-umbilical point on a surface, while they may or may not exist about an umbilic. We show asymptotic coordinates exist in some special cases, e.g., about a point of a surface which has a neighborhood on which the curvature is a negative constant, and about a non-umbilical point on a negative constant, and about a non-umbilical point on a minimal surface (problem 88).

THEOREM (Gauss 1827). *Let γ be an arbitrary univalent curve in M^2 parameterized by arc length on (a, b), let X be the (unit) tangent to γ, and let Y be a unit C^∞ field along γ such that $\langle X, Y\rangle = 0$. Then the Fermi coordinate system induced by Y on a neighborhood of γ is an orthogonal coordinate system about γ which is called a set of "geodesic parallel coordinates." This proves the existence of orthogonal coordinates about any point on a two-dimensional Riemannian manifold.*

Proof. Let ϕ be the Fermi coordinate map from the neighborhood U of γ onto the set V in R^2.

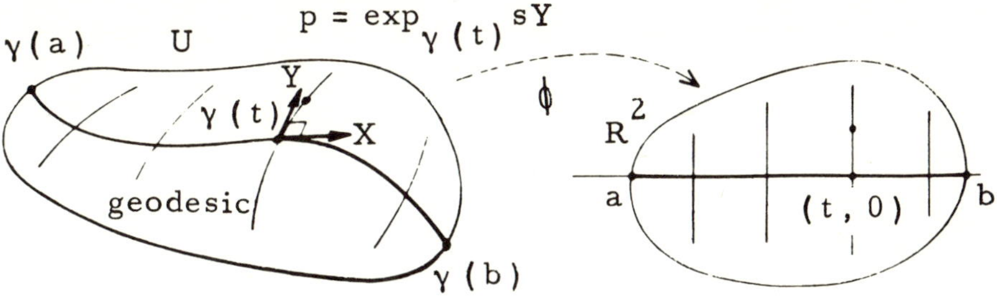

Fig. 9.1 Fermi Cordinates

Then for (t, s) in V, $\phi^{-1}(t, s) = \exp_{\gamma(t)} sY$. We let X and Y be the coordinate fields on U which extend X and Y along γ. Since the y-curves are geodesics parameterized by arc length, $D_Y Y = 0$ and $\langle Y, Y\rangle = 1$, where D is the Riemannian connexion. We compute $Y\langle X, Y\rangle = \langle D_Y X, Y\rangle + \langle X, D_Y Y\rangle = \langle D_Y X, Y\rangle = 1/2X\langle Y, Y\rangle = 0$, since the torsion is zero, so $D_Y X - D_X Y = [Y, X] = 0$. Thus $\langle X, Y\rangle$ is constant along the y-curves and since $\langle X, Y\rangle = 0$ on γ we have $\langle X, Y\rangle = 0$ on U.//

One way to paraphrase the above situation is to say "if segments of equal length (lying in U) are laid off along geodesics that are orthogonal to a univalent curve γ, then their endpoints determine an orthogonal trajectory to the family of geodesics."

THEOREM. *If m is a non-umbilical point on a surface M in R^3, then there exists a set of principal coordinates in a neighborhood U of m.*

Proof. Since m is non-umbilic, there is a neighborhood V of m which contains no umbilics. Assume V is oriented via a unit field N, and let $L(X) = \bar{D}_X N$ as usual, where $\bar{D}$ is the Riemannian connexion on R^3. Let X and Y be C^∞ orthonormal principal vector fields on V with $L(X) = kX$, $L(Y) = hY$, and $k < h$, which corresponds to the notation of Chapter 3. We seek non-vanishing C^∞ functions f and g defined on a neighborhood of m such that the fields $Z = fX$ and $W = gY$ satisfy the condition $[Z, W] = 0$. Finding f and g, we can apply the theorem of section 1; to obtain the desired principal coordinates.

We compute $[fX, gY] = f(Xg)Y - g(Yf)X + fg(aX - bY)$, where $a = (Yk)/(h - k)$ and $b = -(Xh)/(h - k)$ by theorem 3.2. Hence $[Z, W] = 0$ if $(Xg) - bg = 0$ and $(Yf) - af = 0$. Thus we may prescribe $g = 1$ on the integral curve of Y through m, and then on each integral curve $\gamma(t)$ of X we have the differential equation

$$\frac{dg \circ \gamma(t)}{dt} - (b \circ \gamma)(t)(g \circ \gamma)(t) = 0.$$

From the existence theory of ordinary differential equations we get g defined and C^∞ on a neighborhood of m with $g > 0$. Similarly, we obtain f.//

One can write the differential equations $Xg = bg$ and $Yf = af$ as first-order linear partial differential equations in terms of a coordinate system u, v about m. This follows, since $X = b_1(\partial/\partial u) + b_2(\partial/\partial v)$ and $Y = a_1(\partial/\partial u) + a_2(\partial/\partial v)$ defines C^∞ functions a_i and b_i, and then one must solve,

$$b_1 \frac{\partial g}{\partial u} + b_2 \frac{\partial g}{\partial v} = bg \quad \text{and} \quad a_1 \frac{\partial f}{\partial u} + a_2 \frac{\partial f}{\partial v} = af.$$

THEOREM. *If m is contained in the neighborhood U on a surface with constant $K = -a^2 < 0$ on U, then there exists a set of asymptotic coordinates about m.*

Proof. Let X and Y be orthonormal principal fields on U with $LX = kX$ and $LY = hY$, $k < 0 < h$. Let $b = (a^2 + k^2)^{-1/2}$, $Z = b(aX - kY)$, and $W = b(-aX - kY)$. Then $\langle LZ, Z\rangle = \langle LW, W\rangle = 0$, and Z and W are clearly independent. Using theorem 3.2, one computes $[Z, W] = 0$. Hence, the desired coordinates exists.//

Section 9.6. Isothermal coordinates and Riemann surfaces.

The principal reference for this section is Samelson. Let M be a Riemannian 2-manifold.

Let x, y be an arbitrary coordinate system on a neighborhood U of M. We seek functions f and g so the map $p \to (f(p), g(p))$ will define a conformal coordinate system about m in U. If f and g exist, let $E = \langle\partial/\partial f, \partial/\partial f\rangle$, $F = \langle\partial/\partial f, \partial/\partial g\rangle$, and $G = \langle\partial/\partial g, \partial/\partial g\rangle$. Then grad f

$= (1/W^2)(G\partial/\partial f - F\partial/\partial g)$ where $W = (EG - F^2)^{1/2}$. If f and g are orthogonal coordinates, then $F = 0$. If they are also conformal coordinates, then $E = G$ and $|\text{grad } f|^2 = (1/E) = |\text{grad } g|^2$. Thus coordinates f and g are conformal iff $\langle \text{grad } f, \text{grad } g\rangle = 0$ and $|\text{grad } f|^2 = |\text{grad } g|^2$.

In terms of the x, y coordinate system, $\langle \text{grad } f, \text{grad } g\rangle = g_x(Gf_x - Ff_y) - g_y(Ff_x - Ef_y)$ where $g_x = \partial g/\partial x$, etc., and E, F, and G now belong to x and y, i.e., $E = \langle \partial/\partial x, \partial/\partial x\rangle$, etc. Thus $\langle \text{grad } f, \text{grad } g\rangle = 0$ if there is a function ρ on U with

$$(1) \qquad g_x = \rho(Ff_x - Ef_y) \quad \text{and} \quad g_y = \rho(Gf_x - Ff_y).$$

Then $|\text{grad } g|^2 = \rho^2 W^2 |\text{grad } f|^2$, so let $\rho = 1/W$. The equations (1) become a generalization of the Cauchy-Riemann equations. For a particular f, one can solve the system (1) for g iff $g_{xy} = g_{yx}$ or

$$(2) \qquad \frac{\partial}{\partial x}\left[\frac{Gf_x - Ff_y}{\sqrt{EG - F^2}}\right] + \frac{\partial}{\partial y}\left[\frac{Ef_y - Ff_x}{\sqrt{EG - F^2}}\right] = 0.$$

Equation (2) is the classical *Beltrami equation*, a generalized form of the Laplace equation. Indeed, the left side of (2) is $W\Delta f$. Classically, $\langle \text{grad } f, \text{grad } g\rangle$ is called the first *Beltrami operator* on f and g and the Laplacian Δ is called the *second Beltrami operator*.

The theory of elliptic partial differential equations gives the existence of non-trivial solutions of (2) about a point in U which proves the following theorem.

THEOREM. *There exists a system of isothormal (conformal) coordinates about any point of Riemannian 2-manifold.*

On manifolds M as described in this theorem, if we restrict ourselves to conformal coordinate systems then, when the domains of these coordinate systems intersect, they induce a conformal map from one open set of R^2 onto another. Since R^2 is the underlying set for the space of complex numbers, these conformal maps must be given by analytic functions from one open set of C onto another. Thus at each point m of M we have diffeos of a neighborhood of m onto an open set in C which are related by analytic functions on the intersection of their domains. When M is covered by neighborhoods *such that the analytic functions induced by overlapping neighborhoods are orientation preserving, then M is called a Riemann surface and the study of these objects leads to a rich theory (see Ahlfors and Sario).*

Problems.

79. Let T be a C^∞ vector field on the Riemannian manifold M and define A_T: $M_m \to M_m$ by $A_T(X) = D_X T$, where D is the Riemannian connexion. Show that div$T$ = trace A_T. Show A_T is self-adjoint iff $d \circ G(T) = 0$ (T is *closed*). Let $(T^\perp)_m = [X$ in M_m: $\langle X, T_m \rangle = 0]$. If T is closed, show $T^\perp$ is an integrable $(n-1)$-dim distribution on the subset of M where $T \neq 0$.

80. (Frobenius) Let $w_1,\ldots,w_k$ be a set of independent C^∞ 1-forms on a C^∞ n-manifold M with $k < n$. Define an $(n-k)$-dim distribution P on M by $P_m = [X$ in M_m: $w_i(X) = 0$ for $i = 1,\ldots,k]$. Show that P is integrable iff $dw_i = \Sigma_{1 \leq r < s \leq k} a_{irs} w_r \wedge w_s$ for all i. (For generalizations of this result, see Kuranishi or Johnson.)

81. If $G = GL(n, R)$, I is the identity in G, A in G_I, and σ: $t \to e^{tA} \equiv I + tA + (tA)^2/2! + \ldots + (tA)^n/n! + \ldots$, show $\sigma(t)$ is a 1-parameter subgroup of G with tangent A at $t = 0$. Thus show $e^{tA} = \exp_I tA$ for all t (see problem 46).

82. Show the map $(m, X) \to |X|$ is C^∞ on the set $N = [(m, X)$ in $T(M)$: $X \neq 0]$.

83. If M is a Riemannian manifold and A is a compact set in M, show there exists a real number $r > 0$ such that the ball $B(m, r)$ is convex for all m in A.

84. If G is a Lie group, g in G, X in the Lie algebra, and $g = \exp X$, show that $h^2 = g$ where $h = \exp(X/2)$. if h in $SL(2, R) = [g$ in $GL(2, R)$: det $g = 1]$, show the trace $(h^2) \geq -2$. Use this to prove the exp map is not always onto even when the connexion is complete.

85. Let D be a connexion on M. Show the curvature $R \equiv 0$ iff the horizontal distribution H on $B(M)$ is integrable (section 5.5). Show that $R \equiv 0$ implies parallel translation is independent of the path (problem 45).

86. Show there exists at least one umbilic on any compact convex C^∞ surface in R^3. (It was conjectured by Caratheodory, and proven by Bol and Hamburger independently, that a compact convex surface has at least two umbilics.)

87. If M is a surface in R^3, U a coordinate domain on M with coordinate fields X and Y, show the area of U is equal to $\int_U(\langle X, X\rangle\langle Y, Y\rangle - \langle X, Y\rangle^2)^{1/2}$. Let f be in $C^\infty(U, R)$ and define a *normal deformation* belonging to f by $\phi_t(p) = p + tf(p)N_p$ for p in U and N a C^∞ unit normal on U. Let $J(t)$ be the area of $\phi_t(U)$. Show $J'(0) = 0$ for all f iff U is a minimal surface ($H \equiv 0$):

88. Show that about any non-umbilic point on a minimal surface there exists an isothermal coordinate system x, y whose coordinate curves are lines of curvature. Show the functions $z = (x + y)/2$ and $w = (x - y)/2$ define an isothermal coordinate system whose coordinate curves are asymptotic curves which bisect the x, y coordinate curves.

89. Using the notation of section 3.4, let u, v be conformal coordinates on domain B with $E = G = \langle T_u, T_u\rangle$. Show $T_{uu} + T_{vv} = -HGN$. If f is C^∞ on B, show $\Delta f = (1/G)(f_{uu} + f_{vv})$. Let I: $M \to R^3$ be the inclusion map of a surface M into R^3, and let $x_i = u_i \circ I$ on M for $i = 1, 2, 3$. Defining $\Delta I = (\Delta x_1, \Delta x_2, \Delta x_3)$, show $\Delta I = -HN$ on B. Thus if M is minimal, then the functions x_i are harmonic on M.

90. Let f_1, f_2, and f_3 be three analytic functions defined on an open set B in the complex numbers C. Let Z: $B \to C^3$ by $Z(w) = (f_1(w), f_2(w), f_3(w))$ and define X and Y mapping B into R^3 by $X = ReZ$ and $Y = ImZ$ so $Z = X + iY$. If $Z' \bullet Z' = 0$ and $X_u \bullet X_u > 0$ on B, show the maps X and Y each define an immersion of B into R^3 whose image locally is a minimal surface. Conversely, if M a minimal surface in R^3 and m in M, show there is an open set B in C and analytic functions f_1, f_2, and f_3 defined on B such that $X(B)$ is a neighborhood of m in M.

91. A *Weingarten* surface M in R^3 is a surface whose principal curvatures are functionally dependent. Let W: $M \to R^2$ by

$W(m) = (k(m), h(m))$, where $k \leq h$, and call the image of W the W-diagram. Show there exists no compact Weingarten surface of positive Gauss curvature whose W-diagram has negative slope (see section 3.1). Show a compact surface with $K > 0$ and H constant is a sphere. Hopf[3] has shown a compact surface with (a) constant mean curvature and (b) Euler characteristic zero, is a sphere. It is an open question whether the assumption (b) can be dropped.

92. Let X and Y be the coordinate fields for a set of orthogonal coordinates on a surface. Show there exist conformal coordinate with the same coordinate curves (as images) iff $YX[\log (E/G)] = 0$.

10. Topics in Riemannian Geometry

Section 10.1. Jacobi fields and conjugate points.

In order to study the minimizing properties of geodesics, we study one and two parameter families of curves and the vector fields which they induce. Our main tools are developed in the following three propositions.

Let Q and M be C^∞ manifolds, and let f be a C^∞ map of Q into M. A *$T(M)$-valued vector field on Q associated with f*, or a *$T(M)_f$ field on Q*, is a C^∞ function A from Q into $T(M)$, the tangent bundle to M, such that $A(p)$ lies in $M_{f(p)}$ for all p in Q. The field A is a *tangent* $T(M)_f$ field on Q if $A = f_* A'$ for some C^∞ field A' on Q.

For the remainder of this section, let Q, M, and f be as just decribed, and let D denote a connexion on M.

If A and Z are $T(M)_f$ fields on Q and $A = f_* A'$ is tangent, then we can define $D_A Z$ to be a $T(M)_f$ field on Q. This is possible, since for a particular p in Q the field Z gives a well-defined C^∞ field along a curve through $f(p)$ with tangent A_p. More explicitly, let $y_1, \ldots, y_n$ be a coordinate system on M about $f(p)$ and let $Y_i = \partial/\partial y_i$. Let U be an open set about p such that $f(U)$ is contained in the domain of the y_i. Then $Z = \Sigma_1^n z_i Y_i$ defines real valued C^∞ functions z_i on U and

(1) $$D_A Z = \Sigma_1^n [(A' z_i) Y_i + z_i (D_A Y_i)]$$

on U. Letting equation (1) define $D_A Z$ on U, we leave it to the reader to show this definition is independent of the coordinate system. Notice that $D_A Z$ is *not* necessarily a tangent $T(M)_f$ field even when both A and Z are tangent.

If A and B are tangent $T(M)_f$ fields on Q, then we define the tangent $T(M)_f$ field $[A, B]$ by $[A, B](p) = f_*([A', B']_p)$ where $A = f_*A'$, $B = f_*B'$ and p in Q.

Proposition 1. Let $A, B, X, {}^{\#}Z$ be $T(M)_f$ fields on Q, let A and B be tangent, and let g be a real-valued C^∞ function on Q. Then the following equations are valid:

(2) $D_{(gA)}X = g(D_A X)$

(3) $D_A(gX) = (A'g)X + g(D_A X)$

(4) $D_{(A+B)}X = D_A X + D_B X$

(5) $D_A(X + Z) = D_A X + D_A Z.$

Proof. All four equations follow in a straightforward way from the definition (1) and the standard properties for D.//

Observe now for $T(M)_f$ fields X and Z we can define the $T(M)_f$ field Tor (X, Z) by $[\text{Tor}\,(X, Z)](p) = \text{Tor}\,(X_p, Z_p)$ since Tor is a tensor; moreover, the linear transformation-valued tensor $[R(X, Z)](p) = R(X_p, Z_p)$ is defined by p in Q.

Proposition 2. With the hypothesis of proposition 1, the following equations are valid:

(6) $\text{Tor}\,(A, B) = D_A B - D_B A - [A, B]$

(7) $R(A, B)X = D_A D_B X - D_B D_A X - D_{[A,B]}X.$

Proof. Using the notation developed above for equation (1), let $A = \Sigma_1^n a_i Y_i$ and $B = \Sigma_1^n b_j Y_j$. Then on U, Tor $(A, B) = \Sigma_{i,j} a_i b_j$ Tor $(Y_i, Y_j) = \Sigma_{i,j} a_i b_j (D_{Y_i} Y_j - D_{Y_j} Y_i)$, but $[A, B] = \Sigma_j (A'b_j)Y_j - \Sigma_i (B'a_i)Y_i$

and $D_A B - D_B A = \Sigma_j (A' b_j) Y_j + \Sigma_{i,j} b_j a_i D_{Y_i} Y_j - \Sigma_i (B' a_i) Y_i - \Sigma_{i,j} a_i b_j D_{Y_j} Y_i$; hence, equation (6) follows.

A similar computation gives (7).//

Proposition 3. If M is a Riemannian manifold and D is the Riemannian connexion, then with the hypothesis of proposition 1, the following equations are valid:

$$A'\langle X, Z\rangle = \langle D_A X, Z\rangle + \langle X, D_A Z\rangle \tag{8}$$

$$\text{Tor}\,(X, Z) = 0. \tag{9}$$

Proof. Since Tor = 0 in this case, equation (9) is trivial.

To verify (8), let $Y_1, \ldots, Y_n$ be an orthonormal base field with no loss of generality. Letting $X = \Sigma_1^n x_i Y_i$ and $Z = \Sigma_1^n z_j Y_j$, we have $A'\langle X, Z\rangle = A'(\Sigma_1^n x_i z_i) = \Sigma_1{}^n[(A' x_i) z_i + x_i (A' z_i)]$, while $\langle D_A X, Z\rangle + \langle X, D_A Z\rangle = \Sigma_i (A' x_i) z_i + \Sigma_{i,j} x_i z_j \langle D_A Y_i, Y_j\rangle + \Sigma_j x_j (A' z_j) + \Sigma_{i,j} x_i z_j \langle Y_i, D_A Y_j\rangle$. But $\langle D_A Y_i, Y_j\rangle + \langle Y_i, D_A Y_j\rangle = A\langle Y_i, Y_j\rangle = 0$; hence (8) follows.//

We specialize and let Q be an open set in R^2. For convenience, let t and w be the first and second coordinate functions, respectively, on R^2; then $T = f_*(\partial/\partial t)$ and $W = f_*(\partial/\partial w)$ are tangent $T(M)_f$ fields on Q. Moreover, assume the t-varying curves obtained from f by holding w constant are geodesics with respect to a connexion D on M; thus $D_T T \equiv 0$ on Q. When f and Q satisfy the conditions of the above three sentences, we call f a *one-parameter family of geodesics*. When we only assume Q is an open subset of R^2, we call f a *one-parameter family of curves*.

THEOREM 1. *If f is a one-parameter family of geodesics on Q and D is torsion free, then $D_T^2 W = R(T, W)T$ on Q.*

Proof. Since $[T, W] = 0$ and Tor = 0, we have $D_T W = D_W T$. Hence $D_T^2 W = D_T(D_T W) = D_T(D_W T) = D_W(D_T T) + R(T, W)T = R(T, W)T$ by (6) and (7) and the fact $D_T T = 0$.//

Let T be the tangent field along a geodesic for a torsion-free connexion D on M. Then a C^∞ field Z along the goedesic is a *Jacobi field* if $D_T^2 Z = R(T, Z)T$. Notice the set of Jacobi fields along a geodesic is a vector space over the real field from the linearity of the defining condition.

THEOREM 2. *A Jacobi field Z along a geodesic is uniquely determined by its value and the value of $D_T Z$ at one point on the geodesic.*

Proof. Let $e_1,\ldots,e_n = T$ be a parallel base along the geodesic so $Z(t) = \sum_{i=1}^n z_i(t)e_i$ where t is the parameter on the geodesic and z_i are C^∞ real-valued functions. Then $D_T Z = \Sigma_i z_i' e_i$ and $D_T^2 Z = \Sigma_i z_i'' e_i$. Letting $R(e_i, e_j)e_r = \Sigma_{k=1}^n R_{ijrk} e_k$, we have $R(T, Z)T = R(e_n, \Sigma z_j e_j)e_n = \Sigma_{j,k} z_j R_{njnk} e_k$. Hence Z is a Jacobi field iff $z_k'' = \Sigma_{j=1}^n z_j R_{njnk}$ for all k. The conclusion of the theorem now follows from the uniqueness theorem for solutions of second-order differential equations.//

Corollary. The vector space of Jacobi fields along a geodesic has finite dimension equal to $2n$. The subspace of Jacobi fields along a geodesic that vanish at a fixed point has dimension n.

The two theorems above indicate two ways of obtaining Jacobi fields, e.g., use Theorem 2 and existence theory from differential equations or use Theorem 1 by finding a one-parameter family of geodesics. We now illustrate the latter procedure.

We first fix some notation. For any vector A in the tangent space M_m we let A' be the naturally associated "constant" vector field on M_m. We use the notation of section 9.3, for a point X in M_m, $A'_X = \eta_X(A)$; or if $e_1,\ldots,e_n$ a base of M_m and $w_1,\ldots,w_n$ its dual base with $A = \Sigma_1^n a_i e_i$, then $A' = \Sigma_1^n a_i(\partial/\partial w_i)$.

THEOREM 3. *Let X and A be any vectors in M_m. Let $Q = [(t, w)$ in R^2: $\exp_m$ is defined on $t(X + wA)]$, which is an open set in R^2. Let f: $Q \to M$ by $f(t, w) = \exp_m t(X + wA)$. Then f is a one-parameter family of geodesics and $(\exp_m)_*(tA')$ is a Jacobi field along each geodesic.*

Proof. That f is a one-parameter family of geodesic follows from the definition of the exponential map, i.e., $\exp_m$ maps rays in M_m into geodesics emanating from m. Then $W = (\exp_m)_*(tA')$ is a Jacobi field by Theorem 1 (see Fig. 10.1).

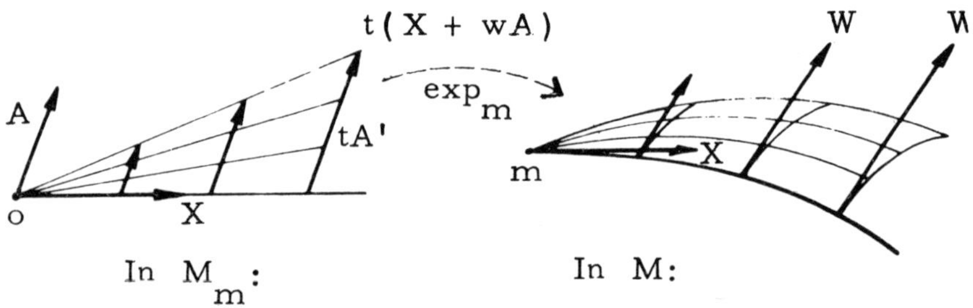

Fig. 10.1 Jacobi Field

A point X in M_m is a *conjugate point* if $\exp_m$ is singular at X. The point (m, X) in $T(M)$ is called a *conjugate point* if X is a conjugate point in M_m. *A point m in M is conjugate to a point p in M along a geodesic g* if there is a conjugate point X in M_m such that $\exp_m X = p$ and g is a reparameterization of the geodesic $g_X(t) = \exp_m tX$.

Notice there is always a neighborhood of zero in M_m that is free of conjugate points since $(\exp_m)_*$ is non-singular at zero (section 9.3, Cor. 1). For a trivial (and too special) example of conjugate points, let M be the unit sphere about the origin in R^3. Then the south pole is conjugate to the north pole along any geodesic (great circle); moreover, the north pole is conjugate to itself along any geodesic. To see this let p be the north pole, then $\exp_p$ is completely singular on circles about zero in M_p which have radius $k\pi$ for integral k.

THEOREM 4. *A point X in M_m is a conjugate point iff there is a non-trivial Jacobi field along g_X that vanishes at m and $\exp_m X$.*

Proof. If $\exp_m$ is singular at X let $A' \neq 0$ be a vector such that $(\exp_m)_* A' = 0$. Then, letting A' denote the associated constant vector field on M_m, the field $(\exp_m)_* tA'$ is a non-trivial Jacobi field along g_X that vanishes at $m(t = 0)$ and $\exp_m X(t = 1)$.

Conversely, let Z be a non-trivial Jacobi field along g_X with $Z(0) = Z(1) = 0$. Let $A = D_X Z$ in M_m and let A' be the associated constant field on M_m. Let $Z' = (\exp_m)_*(tA')$. Then $D_X Z' = D_X[t(\exp_m)_* A'] = (\exp_m)_* A' + tD_X[(\exp_m)_* A']$, and at $t = 0$, $D_X Z' = A$ since at zero $(\exp_m)_* A'_0 = A$. Thus by uniqueness (Theorem 2) $Z = Z'$, and hence $Z'(1) = (\exp_m)_* A'_X = 0$. Since Z is non-trivial $A' \neq 0$ and thus $\exp_m$ is singular at X.//

Corollary. A point m is conjugate to a point p along a geodesic g iff p is conjugate to m along g.

THEOREM 5. *Let g be a geodesic whose parameter domain includes $[b, c]$ and suppose $g(b)$ is not conjugate to $g(c)$ along g. Then there is a unique Jacobi field Z along g with prescribed values at $g(b)$ and $g(c)$.*

Proof. Suppose $Z(b)$ and $Z(c)$ are given. By hypothesis, the map $\exp_{g(b)}$ is non-singular at the point X in $M_{g(b)}$ where $\exp_{g(b)} X = g(c)$, i.e., $g(t) = \exp_{g(b)}(\frac{t - b}{c - b})X$; hence there is a unique vector A' such that $(\exp_{g(b)})_* A' = Z(c)$. Let $Z_1 = (\exp_{g(b)})_*(tA')$ along $\exp_{g(b)} tX$ (which is along g). Similarly, we get a unique vector B' tangent to $M_{g(c)}$ such that $(\exp_{g(c)})_* B' = Z(b)$. Let $Z_2 = (\exp_{g(c)})_*(tB')$ along $\exp_{g(c)} tY$ where $\exp_{g(c)} Y = g(b)$. Then $Z = Z_1 + Z_2$ is a Jacobi field along g with the required values at $g(b)$ and $g(c)$. Furthermore Z is unique, for if W where another such field, then $Z - W$ would be a Jacobi field that vanishes at $g(b)$ and $g(c)$ and hence must be trivial, so $Z = W$.//

Section 10.2. *First and second variation formulae.*

Throughout this section let M be a C^∞ Riemannian n-manifold which is Hausdorff, and let D be the Riemannian connexion. For an alternate approach to the material of this section see Ambrose.[3]

THEOREM.6. *Let f be a one-parameter family of geodesics in M which are parameterized by arc length. Then $\langle W\ T\rangle$ is constant along* *each geodesic.*

Proof. The function $<T, T> = 1$ on the domain of f; hence, $0 = W'<T, T> = 2<D_W T, T>$ by proposition 3. Thus $T<W, T> = <D_T W, T> + <W, D_T T> = <D_W T, T> = 0$, since $D_T T = 0$.//

Corollary ("perpendicular lemma"). Let X be a unit vector in M_m. Let A be in M_m with $<A, X> = 0$ and let A' be the associated constant vector field on M_m. Then $(\exp_m)_* A'$ is perpendicular to the geodesic g_X at all points where g_X defined.

Proof. We may assume A is a unit vector and then define $f(t, w) = \exp_m t[(\cos w)X + (\sin w)A]$ for t in the domain of g_X and w in an interval about zero. Then f is a one-parameter family of geodesics which are parameterized by arc length. Applying the above theorem, we have $<W, T>$ constant along each geodesic. In this case, $W = (\exp_m)_* t[-(\sin w)X + (\cos w)A]$ and $w = 0$ along g_X, hence, $<(\exp_m)_* tA, T> = t<(\exp_m)_* A, T>$ is constant along g_X. This vanishes at $t = 0$, so $<(\exp_m)_* A, T> = 0$ along g_X.//

Let f be a one-parameter family of curves with domain Q and assume Q contains the set $(t, 0)$ for $0 \leq t \leq b$. Let $f_w(t) = f(t, w)$ for (t, w) in Q, and let $L(w)$ be the length of the curve f_w on $[0, b]$, i.e., $L(w) = \int_0^b \sqrt{<T, T>}dt$. We define the *first and second variations of L in the direction f* to be the numbers $L'(0)$ and $L''(0)$, respectively, where $L' = dL/dw$. Actually, we should call $L'(0)$ the "first derivative of L in the direction of the variation f evaluated at f_0 on $[0, b]$," and a similar statement should be made for the "second variation." Henceforth we refer to f_0 as the *base curve.*

THEOREM 7. *In terms of the notation just developed.*

$$L'(0) = <W, T>\Big|_{(0,\,0)}^{(b,\,0)} - \int_0^b <W, D_T T>_{w=0}dt$$

when f_0 is parameterized by arc length. Thus if f_0 is a geodesic, then

$$L'(0) = <W, T>\Big|_{(0,\,0)}^{(b,\,0)}.$$

Proof. We compute,

$$L'(w) = \int_0^b (\partial/\partial w)\sqrt{<T, T>}dt = \int_0^b <T, T>^{-(1/2)}<D_W T, T>dt.$$

When $w = 0$, $<T, T> = 1$ and

$$<D_W T, T> = <D_T W, T> = \frac{d}{dt}<W, T> - <W, D_T T>$$

which we integrate to obtain the above formula.//

Notice that theorem 7 shows $L'(0)$ only depends on the vector field W along the base curve f_0 and we may use the general formula of theorem 7 to define the *first variation of L in the direction of the field W* where W is any C^∞ field on the base curve. For each such C^∞ field W on a base curve σ we can define a one-parameter family f such that $W = f_*(\partial/\partial w)$ by letting $f(t, w) = \exp_{\sigma(t)}(wW_{\sigma(t)})$.

A curve σ between points p and q in M is called an *extremal to the fixed end-point problem* if $L'(0) = 0$ for every one-parameter family of curves f such that $f_0 = \sigma$ on $[0, b]$ and $f(0, w) = p$, while $f(b, w) = q$ for w near 0.

THEOREM 8. *A curve σ between points p and q in M is an extremal iff it is a geodesic.*

Proof. If σ is a geodesic and the end-points are fixed so $W = 0$ at p and q, then $L'(0) = 0$ by theorem 7.

Conversely, if $L'(0) = 0$ and $W = 0$ at p and q, then $\int_0^b <W, D_T T>dt = 0$ for all W belonging to admissable (fixed end-point) one-parameter variations f of σ. If at some point m on σ between p and q we suppose $(D_T T)_m \neq 0$, then let $W = hD_T T$ where h is a C^∞ "bump" function such that $h(m) = 1$, $h \geq 0$, and $h = 0$ outside a neighborhood of m on which $D_T T$ doesn't vanish. By the remarks after theorem 7, there is a one-parameter family f belonging to W. In this case $<W, D_T T> = h<D_T T, D_T T> \geq 0$ is a non-negative function which is non-zero on a neighborhood of t' where $\sigma(t') = m$, hence $\int_0^b <W, D_T T>dt > 0$, which is a contradiction. Thus $D_T T = 0$, and σ is a geodesic.//

THEOREM 9. *For a point m in M, let $r > 0$ be chosen so $\exp_m$ maps the set $\hat{B} = [X \text{ in } M_m\colon |X| < r]$ diffeomorphically onto its image B. Then B is the metric ball $B(m, r) = [p \text{ in } M\colon d(m, p) < r]$. Furthermore, if X in $\hat{B}$ and $p = \exp_m X$ then $d(m, p) = |X|$, and the geodesic $g_X(t) = \exp_m tX$ defined on $[0, 1]$, realizes the absolute minimum possible curve-length from m to p.*

Proof. If T is the tangent to g_X, then $\langle T, T\rangle$ is constant on g_X so $|g_X|_0^1 = |X|$. We must show any other broken C^∞ curve σ from m to p has a length which is greater than or equal to $|X|$, and the theorem will follow.

First suppose σ is defined on $[0, b]$ and $\sigma(t)$ is in B for all t in $[0, b]$. Furthermore, suppose σ never returns to m after $t = 0$, or we could obviously obtain a shorter curve from m to p. Let $\exp = \exp_m$ and let $\exp^{-1}$ be the inverse map of $\exp|_{\hat{B}}$. Let $f(t) = |\exp^{-1}\sigma(t)|$ for t in $[0, b]$, which defines a broken C^∞ function f. Let $\bar{\sigma}(t) = \exp^{-1}\sigma(t)$, $\bar{\gamma}(t) = f(t)X/|X|$, and $\gamma(t) = \exp \bar{\gamma}(t)$. Thus γ is a reparameterization of g_X which has the same "radial velocity" as σ. Decompose the tangent to $\bar{\sigma}$ into a radial component A and a vector V which is orthogonal to A, thus $T_{\bar{\sigma}} = A + V$ on $[0, b]$, (actually, $A(t) = f'(t)\bar{\sigma}(t)/f(t)$ for $t > 0$). Using the perpendicular lemma proved above, we know $\exp_* A$ is perpendicular to $\exp_* V$, so $|T_\sigma| = |\exp_* A + \exp_* V| \geq |\exp_* A| = |T_\gamma|$. Hence, $|\sigma|_0^b \geq |\gamma|_0^b$. Since γ is a reparameterization of g_X, we have $|\gamma|_0^b \geq |g_X|_0^1 = |X|$, where the inequality is strict if f is not an increasing function. Thus, $|\sigma|_0^b \geq |X|$.

If $\sigma(t)$ not in B for all t, then $|\sigma| > r > |X|$ by the above paragraph. Hence, $|X| = d(m, p)$ for X in $\hat{B}$, and the geodesic g_X realizes this minimum.//

THEOREM 10. *Let f be a one-parameter family of curves such that the base curve is a geodesic g parameterized by arc length on the interval $[0, b]$. Then $L''(0) =$*

$$\langle D_W W, T\rangle \Big|_{(0,0)}^{(b,0)} + \int_0^b [\langle R(W, T)W, T\rangle + \langle D_T W, D_T W\rangle - (T\langle W, T\rangle)^2]dt.$$

If $\langle W, T\rangle$ is constant along f, then

$$L''(0) = \langle D_W W, T\rangle \Big|_{(0,0)}^{(b,0)} + \int_0^b [\langle R(W, T)W, T\rangle + \langle D_T W, D_T W\rangle]dt.$$

If W a Jacobi field and $<W, T>$ is constant along g, then

$$L''(0) = W<T, W> \Big|_{(0,\,0)}^{(b,\,0)}.$$

Proof. First compute $(\partial/\partial w)\sqrt{<T, T>} = <T, T>^{-1/2}<D_W T, T>$. Then $(\partial^2/\partial w^2)\sqrt{<T, T>} = -<T, T>^{-3/2}<D_W T, T>^2 + <T. T>^{-1/2}(<D_W D_W T, T> + <D_W T, D_W T>)$. Evaluating on $w = 0$, we use $<T, T> = 1$, $D_T T = 0$, and $D_T W = D_W T$, to obtain $(\partial^2/\partial w^2)\sqrt{<T, T>} = <D_W D_T, T> + <D_T W, D_T W> - <D_T W, T>^2 = <R(W, T)W + D_T D_W W, T> + <D_T W, D_T W> - (T<W, T>)^2 = T<D_W W, T> + <R(W, T)W, T> + <D_T W, D_T W> - (T<W, T>)^2$, which gives the first formula for $L''(0)$ by integrating.

If $<W, T>$ is constant along g, then $T<W, T> = 0$ which gives the second formula.

If W is Jacobi, then

$$<R(W, T)W, T> = <R(T, W)T, W> = <D_T^2 W, W> = T<D_T W, W> - <D_T W, D_T W>.$$

Hence $L''(0) = [<D_W W, T> + <D_W T, W>] \Big|_{(0,\,0)}^{(b,\,0)} = W<W, T> \Big|_{(0,\,0)}^{(b,\,0)}.$ //

Notice that the first term is the only term in the above formulae that depends on something more than the vector field W along g.

Corollary. If W vanishes at the end-points of g, then the second variation of L depends only on the field W along g. For any vector field W along g, let $f_W(t, w) = \exp_{g(t)} wW$ be the *natural* one-parameter family associated with W, and then $D_W W = 0$, since the w-varying curves are geodesics. Letting $L''_W(0)$ denote the second variation of L in the direction f_W, then

$$L''_W(0) = \int_0^b [<R(W, T)W, T> + <D_T W, D_T W> - (T<W, T>^2]dt.$$

We next prove two lemmas which are used to prove that geodesics are not minimizing-distance curves past a first conjugate point, and later, to prove conjugate points are isolated along a geodesic in the Riemannian case.

LEMMA 1 (Lagrange identity). *If X and Y are Jacobi fields along a geodesic g with tangent field T, then $\langle D_T X, Y\rangle - \langle X, D_T Y\rangle$ is constant along g.*

Proof. We compute $T(\langle D_T X, Y\rangle - \langle X, D_T Y\rangle) = \langle D_T^2 X, Y\rangle - \langle X, D_T^2 Y\rangle = \langle R(T, X)T, Y\rangle - \langle R(T, Y)T, X\rangle = 0$ by the symmetry of the Riemann-Christoffel curvature tensor.//

LEMMA 2. *Let W be a continuous piecewise C^∞ field along the geodesic g which is parameterized on $[0, b]$, and let $W(0) = 0$. If there is no point $g(t)$ that is conjugate to $g(0)$ for t in $[0, b]$, then*

$$\int_0^b [\langle R(W, T)W, T\rangle + \langle D_T W, D_T W\rangle]dt > \int_0^b [\langle R(Z, T)Z, T\rangle + \langle D_T Z, D_T Z\rangle$$

unless $W = Z$, where Z is the unique Jacobi field along g such that $Z(0) = 0$ and $Z(b) = W(b)$.

Proof. The field Z is well-defined by theorem 5. Let $Z_1,\dots,Z_n$ be a base of $M_{g(b)}$, and extend these vectors by theorem 5 to be Jacobi fields along g that vanish at $g(0)$. Since there is no point $g(t)$ conjugate to $g(0)$, the fields $Z_1,\dots,Z_n$ are a base of $M_{g(t)}$ for all t in $(0, b]$. Using theorem 3, write each $Z_i = tA_i$ where $A_1,\dots,A_n$ are C^∞ fields that are independent on $[0, b]$. Setting $W = \Sigma_{i=1}^n g_i A_i$, we define continuous piece wise C^∞ functions g_i on $[0, b]$. Since $g_i(0) = 0$ we may write $g_i = tf_i$ and thus define continuous piece wise C^∞ functions f_i on $[0, b]$ such that $W = \Sigma f_i Z_i$. Then $Z = \Sigma f_i(b) Z_i$.

Let $D_T W = A + B$ where $A = \Sigma(Tf_i)Z_i$ and $B = \Sigma f_i D_T Z_i$. Then $\langle D_T W, D_T W\rangle = \langle A, A\rangle + 2\langle A, B\rangle + \langle B, B\rangle$, and

$$\langle R(T, W)T, W\rangle = \Sigma f_i \langle R(T, Z_i)T, W\rangle = \Sigma f_i \langle D_T^2 Z_i, W\rangle$$

$$= \Sigma f_i [T\langle D_T Z_i, W\rangle - \langle D_T Z_i, D_T W\rangle]$$

$$= T\langle B, W\rangle - \Sigma(Tf_i)\langle D_T Z_i, W\rangle - \langle B, A\rangle - \langle B, B\rangle.$$

Hence, $\langle R(T, W)T, W\rangle + \langle D_T W, D_T W\rangle = T\langle B, W\rangle + \langle A, A\rangle + \langle A, B\rangle - \Sigma(Tf_i)\langle D_T Z_i, W\rangle$. But

$$\langle A, B\rangle - \Sigma(Tf_i)\langle D_T Z_i, W\rangle = \Sigma(Tf_i)f_j[\langle Z_i D_T Z_j\rangle - \langle D_T Z_i, Z_j\rangle] = 0$$

by the Lagrange identity, since $Z_k(0) = 0$ for all k. Thus $\int_0^b [\langle R(W, T)W, T\rangle + \langle D_T W, D_T W\rangle]dt = \langle B_b, W_b\rangle + \int_0^b \langle A, A\rangle dt$ since W is continuous and $W_0 = 0$. Furthermore, $\langle B_b, W_b\rangle = \langle \Sigma f_i(b)(D_T Z_i)_b, W_b\rangle = \langle (D_T Z)_b, Z_b\rangle = \int_0^b [\langle R(Z, T)Z, T\rangle + \langle D_T Z, D_T Z\rangle]dt$. Since $\int_0^b \langle A, A\rangle dt \geq 0$, the inequality in the conclusion follows unless $A \equiv 0$, which implies f_i are constant so $W \equiv Z$.//

THEOREM 11. *The arc length on a geodesic g does not equal the distance in M beyond the first conjugate point; i.e., if $g(b)$ is the first point of g that is conjugate to $g(0)$, and g is parameterized by arc length, then the distance $d(g(0), g(a)) < a$ for $a > b$.*

Proof. Let Z be a non-trivial Jacobi field along g which vanishes at 0 and b. Then $\langle Z, T\rangle = 0$ by theorem 6 and $L''_Z(0) = 0$ by theorem 10 where L' is computed from the natrual one-parameter family of curves associated with Z. By theorem 9 we obtain $r > 0$, so that the neighborhood $B(g(b), r)$ is the diffeomorphic image of the r-ball about zero in $M_{g(b)}$. Choose numbers a and c such that $0 < c < b < a$ and $g(t)$ is in $B(g(b), r)$ for all t in [c, a]. Thus the interval [c, a] has no pair of points that are conjugate to each other on g. Let Y be the unique Jacobi field along g with $Y(c) = Z(c)$ and $Y(a) = 0$. Let X be the field on $[0, a]$ such that $X(t) = Z(t)$ for t in [0, c] and $X(t) = Y(t)$ for t in [c, a]. Let W be the field on $[0, a]$ such that $W(t) = Z(t)$ for t in $[0, b]$ and $W(t) = 0$ for t in $[b, a]$ (see Fig. 10.2).

Then $L''_W|_0^a = L''_Z|_0^b + L''_W|_b^a = 0$ while $L''_X|_0^a = L''_W|_0^c + L''_Y|_c^a$. By Lemma 2, we have $L''_W|_c^a > L''_Y|_c^a$, which implies $L''_X|_0^a < L''_W|_0^a = 0$. Hence there are broken C^∞ curves in the natural one-parameter family associated with X whose length from $g(0)$ to $g(a)$ is less that a.//

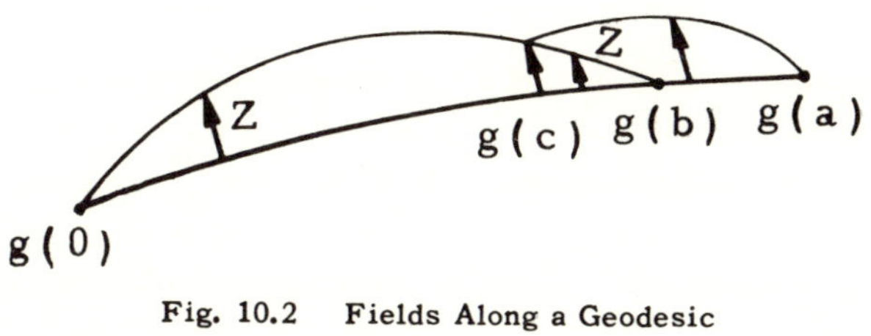

Fig. 10.2 Fields Along a Geodesic

Actually, the arc length on a geodesic may cease to measure distance in M long before a conjugate point is reached (think of a right circular cylinder). The conjugate point is where the geodesic ceases to be a minimum-length curve among *nearby* curves.

THEOREM 12. *The conjugate points of a fixed point on a geodesic occur at isolated values of the parameter.*

Proof. Let $g(b)$ be any point conjugate to $g(0)$ along the geodesic g (notice it is possible that $g(b) = g(0)$). Let $A_1,\dots,A_r$ be a base for the kernel of $(\exp_{g(0)})_*$ at bT_0 in $M_{g(0)}$, where T_t is the tangent to g at $g(t)$, and we assume $\langle T, T\rangle = 1$. Choose $A_{r+1},\dots,A_n$ so $A_1,\dots, A_n$ are independent and let $Z_i(t) = (\exp_{g(0)})_* tA_i$. Then the fields $Z_1,\dots,Z_n$ are Jacobi fields along g that vanish at 0 and are independent for all values of t except 0 and conjugate values. We show there exists an $\epsilon > 0$ such that $Z_1,\dots,Z_n$ are independent for $0 < |t-b| < \epsilon$. This is done by showing $D_TZ_1,\dots,D_TZ_r, Z_{r+1},\dots,Z_n$ are independent at b and then $Z_1/(t-b),\dots,Z_r/(t-b), Z_{r+1},\dots,Z_n$ are independent for $0 < |t-b| < \epsilon$.

Since $A_{r+1},\dots,A_n$ are independent at bT_0, we know $Z_{r+1},\dots,Z_n$ are independent at b. For $i \le r$, $(D_TZ_i)_b \ne 0$, since $(Z_i)_b = 0$ and Z_i is non-trivial. If $\Sigma_{i=1}^r c_i(D_TZ_i)_b = 0$, let $W = \Sigma_1^r c_iZ_i$. Then W is a Jacobi field with $W_b = 0$ and $(D_TW)_b = 0$; hence $W \equiv 0$. For small $a > 0$, we know $Z_1,\dots,Z_r$ are independent, and $\Sigma_1^r c_i(Z_i)_a = 0$ implies $c_i = 0$ for all i. Thus $D_TZ_1,\dots,D_TZ_r$ are independent at b. We now show for $i \le r$ and $j > r$, D_TZ_i is orthogonal to Z_j at b. By the Lagrange identity $\langle D_TZ_i, Z_j\rangle - \langle Z_i, D_TZ_j\rangle$ is constant along g. Since Z_i and Z_j vanish at 0, and Z_i vanishes at b, we have $\langle D_TZ_i, Z_j\rangle = 0$ at b. Thus $D_TZ_1,\dots,D_TZ_r, Z_{r+1},\dots,Z_n$ are independent at b and hence in some neighborhood of b. Since $Z_i(t)/(t-b) \to (D_TZ_i)_b$ as $t \to b$, the conclusion follows.//

Section 10.3. Geometric interpretation of Riemannian curvature.

In this section, let M be a Riemannian manifold, g be a geodesic in M with unit tangent T, A_0 be a unit vector in $M_{g(0)}$ which is orthogonal to T_0, A' be the constant vector field on $M_{g(0)}$ generated by A_0, $\exp = \exp_{g(0)}$, $A = \exp_* A'$, and where $A_t \ne 0$ let $K = \langle R(T, A)A, T\rangle / \langle A, A\rangle$ as a function of t along g. We study the relationship between the Riemannian curvature $K(t)$ of the plane section spanned by A_t and T_t and the length of the vector A_t: The field tA is used in the computation since it is a Jacobi field.

LEMMA. *If* $tA_t \neq 0$, *then*

(1) $T|tA| = \langle D_T tA, tA\rangle/|tA| = |A| + t\langle D_T A, A\rangle/|A|$,

(2) $T^2|tA| = -|tA|K(t) + H(t)$ *where* $H(t) \geq 0$, *and*

(3) $|A_t| = 1 - K(0)(t^2/6) + G(t)t^3$ *for t in a neighborhood of zero where G is* C^∞.

Proof. We compute $T|tA| = T\sqrt{\langle tA, tA\rangle} = \langle D_T tA, tA\rangle/|tA| = \langle A + tD_T A, tA\rangle/|tA| = |A| + t\langle D_T A, A\rangle/|A|$. Thus

$$T^2|tA| = [\langle D_T^2 tA, tA\rangle + \langle D_T tA, D_T tA\rangle - \langle D_T tA, tA\rangle^2/\langle tA, tA\rangle]/|tA|$$

$$= [\langle R(T, tA)T, tA\rangle|tA|^2 + |D_T tA|^2|tA|^2 - \langle D_T tA, tA\rangle^2]/|tA|^3$$

$$= -|tA|K(t) + H(t)$$

where $H(t) = [|D_T tA|^2|tA|^2 - \langle D_T tA, tA\rangle^2]/|tA|^3$. The Schwartz inequality implies $H(t) \geq 0$. A straightforward computation shows as $t \to 0$, $H(t) \to 0$, and $H'(t) \to 0$, since $(D_T A)_0 = 0$ (use normal coord.). Hence as $t \to 0$, we have $|tA| \to 0$, $T|tA| \to |A_0| = 1$, $T^2|tA| \to 0$, and $T^3|tA| \to -K(0)$.

Since A_t does not vanish near $t = 0$, the function $|A_t|$ is C^∞ at 0, and hence $F(t) = |tA_t|$ admits a representation

$$F(t) = F(0) + F'(0)t + F''(0)t^2/2 + F'''(0)t^3/6 + G(t)t^4$$

for t in a neighborhood of 0 where G is a C^∞ function on this neighborhood. Substituting the values for the derivatives of F and cancelling a factor t then gives (3).//

The following theorem derives its form essentially from some class notes of Ambrose.

THEOREM 13. *If $K(t) \leq 0$ for t in $[0, b]$, then $|A_t| \geq |A_0| = 1$ for t in $[0, b]$. Thus if $K \leq 0$ for all plane sections at all points of M, then M has no conjugate points. If $K(0) < 0$, then $|A_t| \geq 1$ for t near zero, and if $K(0) > 0$, then $|A_t| \leq 1$ for t near zero.*

Proof. Let $F(t) = |tA_t| - t|A_0| = |tA_t| - t$. Then $F(0) = 0$, $F'(0) = 0$, and $F''(t) = T^2|tA_t| \geq 0$ if $K(t) \leq 0$. Applying the Mean Value Theorem twice, $F(t) = F'(\bar{t})t = F''(\bar{\bar{t}})\bar{t}t \geq 0$ where $0 \leq \bar{\bar{t}} \leq \bar{t} \leq t \leq b$. Hence $|A_t| \geq 1$ for t in $[0, b]$.

The second sentence of the theorem follows from the first, and the last two sentences follow from (3) in the lemma.//

We obtain a geometric interpretation of Riemannian curvature from the following considerations (see Fig. 10.3). The vector A' at the point bT_0 in $M_{g(0)}$ is tangent to the circle σ of radius b about the origin which lies in the plane of A_0 and T_0. Hence $A = \exp_* A'$ is the tangent at $\exp(bT_0)$ to the curve $\exp \circ \sigma$ in M. If b is sufficiently small, then $\exp \circ \sigma$ passes through points that are exactly b units distant from $g(0)$. If $|A_b| > |A'|$ then the curve $\exp \circ \sigma$ is "stretching" the curve σ near bT_0 and the geodesics emanating from $g(0)$ that are determined by σ are "spreading out." A corresponding statement applies to the case $|A_b| < |A'|$.

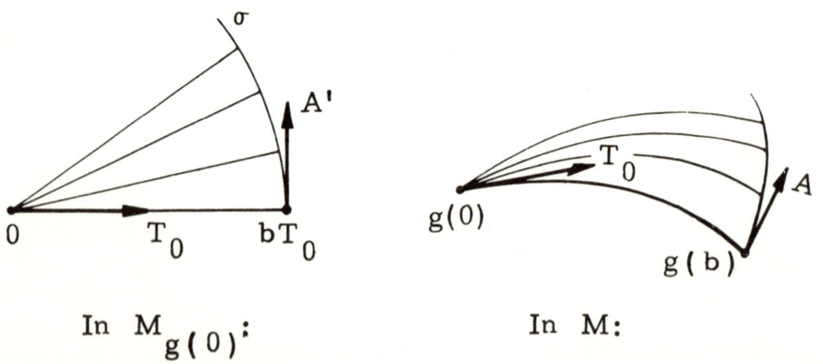

Fig. 10.3 Comparing Geodesics

Section 10.4. *The Morse Index Theorem.*

Our approach to this section is based on the notes of Bott. For further material see Milnor,[3] Ambrose,[2] and Morse. Let M be a C^∞ manifold and let f be a real valued C^∞ function defined on a neighborhood of a point m in M. The point m is a *critical point* of f if $(f_*)_m$ is the zero linear transformation on M_m. If m is a critical point of f, we define a symmetric bilinear function H: $M_m \times M_m \to R$ by $H(X_m, Y_m) = X_m(Yf)$, where Y is any C^∞ vector field about m whose value at m is Y_m. It is a simple exercise to show $H(X_m, Y_m)$ is independent of the field Y and is symmetric and bilinear (see problem 95). The function H is called the *Hessian of f at m*. The *index* of H is defined to be the dimension of a maximal subspace V of M_m on which H is negative definite (and V is maximal if it is not properly contained in a subspace V' on which H is negative definite). The *null space* of H is the subspace $V = [X \text{ in } M_m\text{: } H(X, Y) = 0 \text{ for all } Y \text{ in } M_m]$. The *nullity* of H is the dimension of its null space. We denote the index of H and the nullity of H by $I(f_m)$ and $N(f_m)$, respectively, and call them the *index of f at m* and the *nullity of f at m*, respectively. The *positivity* $P(f_m)$ is the integer such that $P(f_m) + I(f_m) + N(f_m)$ is the dimension of M_m. The index of H intuitively gives the number of dimensions of directions in M_m in which f is decreasing.

Next we need the definition of the conjugate degree of points along a geodesic. Let g be a geodesic in a manifold with connexion. The *conjugate degree of the point* $g(t)$ (with respect to $g(0)$) is the dimension of the kernel of $(\exp_{g(0)})_*$ at tT_0, where T_0 is the unit tangent to g at $g(0)$ and g is parameterized by arc length. Thus the conjugate degree of the point $g(t)$ is the maximum number of linearly independent Jacobi fields along g that vanish at 0 and t.

The Morse Index Theorem relates the concepts just defined. Roughly, it says, for a particular geodesic segment in a Riemannian manifold M, the distance function can be used to define a C^∞ function L on a manifold C, and then the index of L at a particular critical point is equal to the sum of the degrees of conjugate points along the geodesic segment.

For the rest of the section let M be a C^∞ Riemannian Hausdorff n-manifold. If m in M, then a *local geodesic submanifold* of M at m is a submanifold C defined as follows. Let B be an open ball about the origin (zero) in M_m which $\exp_m$ maps diffeomorphically into M, and let V be any subspace of M_m. Then the submanifold $C = [\exp_m X\colon X$ in $B \cap V]$ is a *local geodesic submanifold* of M. Note C contains geodesic segments of geodesics emanating from m whose tangent vectors lie in V (see Fig. 10.4).

LEMMA. *Let A be a convex neighborhood of M, let p_1 and p_2 be in A, let g be the unique geodesic from p_1 to p_2 which lies in A and is parameterized by arc length, let T be the tangent field to g, let C_1 and C_2 be disjoint local geodesic hypersurfaces of A through p_1 and p_2, respectively, that are orthogonal to T, and finally, let $C = C_1 \times C_2$ (see Fig. 10.4). If (m_1, m_2) is a point of C, let $d(m_1, m_2)$ be the distance from m_1 to m_2; thus d is real-valued C^∞ function from C into R (problem 96). Let $W = (W_1, W_2)$ and $U = (U_1, U_2)$ be vectors tangent to C at (p_1, p_2), where W_i and U_i are in M_{p_i} for $i = 1, 2$, and let U also denote the unique Jacobi field along g determined by U_1 and U_2.*

Then $p = (p_1, p_2)$ is a critical point of d on C and

$$H_p(U, W) = U_p(Wd) - [\langle W, D_T U\rangle - II_T(U, W)]_{p_1}^{p_2}$$

where II_T at p_i is the second fundamental form of C_i with respect to the normal in the direction of T.

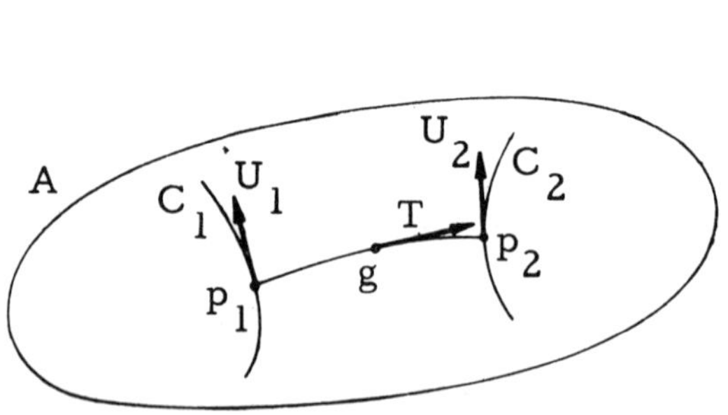

Fig. 10.4 Cross Manifolds

Proof. A *two-parameter family of geodesics* is a C^∞ function f mapping an open set Q in R^3 into M such that the curves $f_{(u_0, w_0)}(t) =$

$f(t, u_0, w_0)$, obtained from f by fixing the coordinates in the last two slots, are geodesics. Let f be such a map and suppose Q contains the set $(t, 0, 0)$ for $0 \leq t \leq b$. Call the geodesic $g = f_{(0,0)}$ the base geodesic and assume g is parameterized by arc length. Let $T = f_*(\partial/\partial t)$, $U = f_*(\partial/\partial u)$, and $W = f_*(\partial/\partial w)$; then T, U, and W are Jacobi fields along the geodesics of f, while $D_T W = D_W T$, $D_T U = D_U T$, and $D_U W = D_W U$ by section 10.1. We assume further that $\langle T, U\rangle$ and $\langle T, W\rangle$ are constant on g; hence $\langle D_T U, T\rangle = 0$ and $\langle D_T W, T\rangle = 0$ on g. For (u, w) near $(0, 0)$, let $L(u, w) = \int_0^b \sqrt{\langle T, T\rangle}dt$. Notice $\langle T, T\rangle$ is a function on Q which depends only on u and w since the t-curves are geodesics. Then $L_w = \partial L/\partial w = \int_0^b \langle T, T\rangle^{-1/2}\langle D_W T, T\rangle dt$, and $(L_w)_{(0,0)} = \int_0^b \langle D_T W, T\rangle dt = 0$ since $\langle T, T\rangle = 1$ on g. Differentiating again,

$$(L_{wu}) = (\partial^2 L/\partial u\partial w) = \int_0^b [-\langle T, T\rangle^{-3/2}\langle D_U T, T\rangle\langle D_W T, T\rangle$$

$$+ \langle T, T\rangle^{-1/2}(\langle D_U D_W T, T\rangle + \langle D_W T, D_U T\rangle)]dt.$$

Evaluating on g, we have

$$(L_{wu})_{(0,0)} = \int_0^b [\langle D_U D_T W, T\rangle + \langle D_T W, D_T U\rangle]dt$$

$$= \int_0^b [\langle R(U, T)W + D_T D_U W, T\rangle + T\langle W, D_T U\rangle - \langle W, D_T^2 U\rangle]dt.$$

But, since U is Jacobi,

$$\langle R(U, T)W, T\rangle - \langle W, D_T^2 U\rangle = \langle R(U, T)W, T\rangle - \langle W, R(T, U)T\rangle = 0;$$

hence,

$$(L_{wu})_{(0,0)} = \int_0^b [\langle D_T D_U W, T\rangle + T\langle W, D_T U\rangle]dt$$

$$= \int_0^b [T\langle D_U W, T\rangle + T\langle W, D_T U\rangle]dt$$

$$= \langle D_U W, T\rangle + \langle W, D_T U\rangle \Big|_{(0, 0, 0)}^{(b, 0, 0)}.$$

We apply the above analysis to prove the lemma. Let $f_{(u,w)}(t) = f(t, u, w)$ be the unique geodesic in A from $\exp_{p_1}(uU_1 + wW_1) = \gamma_1(u, w)$ to $\exp_{p_2}(uU_2 + wW_2) = \gamma_2(u, w)$ which is parameterized on $[0, b]$. Then f is a two-parameter family of geodesics satisfying the above requirements. Furthermore $d(\gamma_1(u, w), \gamma_2(u, w)) = L(u, w)$, hence $H_p(U, W) = \langle D_{U_i} W_i, T\rangle + \langle W_i, D_T U_i\rangle \,|_{i=1}^{i=2}$. But letting D' be the induced Riemannian connexion on C_i, by the Gauss equation we get

$$D_{U_i} W_i = D'_{U_i} W_i - II_T(U_i, W_i)T;$$

hence $H_p(U, W) = [\langle W, D_T U\rangle - II_T(U, W)]_{p_1}^{p_2}$.//

THEOREM 14 (Morse Index Theorem). *Let g be a geodesic in M which is parameterized by arc length on the interval $[0, b]$. Let $r > 0$ be chosen such that the balls $B(g(t), 2r)$ are convex neighborhoods of $g(t)$ for $0 \leq t \leq b$. Let $\bar{m} = (m_1,\ldots,m_k)$ be a sequence of points on g such that $m_i = g(t_i)$, $0 < t_i < t_{i+1} < b$, and*

$$(1) \qquad 0 < d(m_i, m_{i+1}) < r$$

for $i = 0,\ldots,k$ where $m_0 = g(0)$ and $m_{k+1} = g(b)$. Let C_i be a local geodesic submanifold which is orthogonal to g at m_i and contained in $B(m_i, r)$ for $1 \leq i \leq k$, and let $C = C_1 \times \ldots \times C_k$. Let $L\colon\ C \to R$ by $L(\bar{p}) = \Sigma_{i=0}^{k} d(p_i, p_{i+1})$ where $\bar{p} = (p_1,\ldots,p_k)$ in C, $p_0 = g(0)$ and $p_{k+1} = g(b)$ (see Fig. 10.5).

Then L is C^∞ on C, $\bar{m}$ is a critical point of L, the nullity of L at $\bar{m}$ equals the conjugate degree of $g(b)$ (with respect to $g(0)$) and $I(L_{\bar{m}}) = \Sigma_{0 \leq t \leq b} \deg g(t)$.

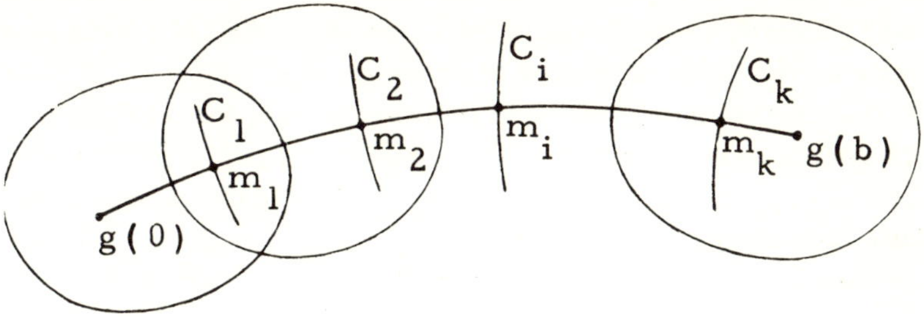

Fig. 10.5 Cross Manifolds

Before proving the theorem we make some remarks. The fact that $N(L_{\overline{m}})$ is the conjugate degree of $g(b)$ is often called the Nullity Theorem. The Index Theorem shows $I(L_{\overline{m}})$ and $N(L_{\overline{m}})$ are independent of the position of the points m_i and the number of points k, as long as condition (1) is satisfied.

Proof. Let L_i: $C \to R$ be defined by $L_i(\overline{p}) = d(p_i, p_{i+1})$ for $i = 0,\ldots,k$. Then L is C^∞ since $L = \Sigma_{i=0}^k L_i$ and each L_i is C^∞. By the lemma, the point $\overline{m}$ is a critical point of each L_i and hence is a critical point of L.

To compute the nullity of L at $\overline{m}$ let U and W be tangent to C at $\overline{m}$ where $U = (U_1;\ldots;U_k)$ and $W = (W_1;\ldots;W_k)$ with U_i and W_i in M_{m_i} for all i. Let $U_0 = W_0$ and $U_{k+1} = W_{k+1}$ be the zero vectors at $g(0)$ and $g(b)$, respectively. By the lemma,

$$U_{\overline{m}}(WL) = \Sigma_0^k U_{\overline{m}}(WL_i)$$

$$= \Sigma_{i=1}^k [\langle W_{i+1}, D_T U_{i+1}^- \rangle - II_T(U_{i+1}, W_{i+1}) - \langle W_i, D_T U_i^+ \rangle + II_T(U_i, W_i)]$$

$$= \Sigma_{i=1}^k \langle W_i, D_T U_i^- - D_T U_i^+ \rangle,$$

where U_i^- is the Jacobi field on $[t_{i-1}, t_i]$ agreeing with U at the endpoints, and $U_i^+ = U_{i+1}^-$. If U is in the null space of H_L at $\overline{m}$, then $U_{\overline{m}}(WL) = 0$ for all W; hence $D_T U_i^- = D_T U_i^+$ for all i, which implies U is a Jacobi field along g that vanishes at 0 and b. This proves the nullity theorem.

We now work on the index of L at $\bar{m}$. Let us refer to a point $(\bar{m}, b)$ in $M^k \times R$ which satisfies the conditions stated in the third sentence of the theorem as an admissable partition. Let $N = k(n-1)$, and for each admissable partition (y, t) let C_y be the product of k local geodesic submanifolds crossing g at the points of y, let $L_{(y,t)}: C_y \to R$ be the function corresponding to L in the theorem, and let F_y map R^N into the tangent space to C_y at y by $F_y(a_1, \ldots, a_N) = (\Sigma_1^{n-1} a_i e_i(y_1); \Sigma_1^{n-1} a_{n-1+j} e_j(y_2), \ldots)$, where $e_1, \ldots, e_{n-1}$, T is an orthonormal parallel base field along g. Then let $H_{(y,t)}$, $I_{(y,t)}$, $P_{(y,t)}$, and $N_{(y,t)}$ denote the Hessian, index, positivity, and nullity respectively, of $H_{L_{(y,t)}} \circ F_y$. Thus $H_{(y,t)}$ is a symmetric bilinear form on R^N which is continuous in y and t.

For each admissable partition (y_0, t_0) there is a neighborhood (in $M^k \times R$) such that

$$(2) \qquad I_{(y,t)} \geq I_{(y_0,t_0)} \quad \text{and} \quad P_{(y,t)} \geq P_{(y_0,t_0)}$$

for (y, t) in this neighborhood. This follows since $I_{(y_0,t_0)}$ is the dimension of a subspace V of R^N such that $H_{(y_0,t_0)}(W, W) < 0$ for all non-zero W in V, and by continuity the inequality must hold on a neighborhood of (y_0, t_0). A similar argument handles the positivity case.

Fix y such that (y, b_1) and (y, b_2) are admissable partitions with $b_1 \leq b_2$. We show

$$(3) \qquad I_{(y,b_1)} \leq I_{(y,b_2)} \quad \text{and} \quad P_{(y,b_1)} \geq P_{(y,b_2)}.$$

For x in the cross manifold C_y, let $A(x) = L_{(y,b_2)}(x)$ and $B(x) = L_{(y,b_1)}(x) + d(g(b_1), g(b_2))$. Then $A(x) \leq B(x)$ by the triangle inequality and $A(y) = B(y)$. On a curve $\gamma(w)$ with tangent W that is tangent to C_y at $y = \gamma(0)$, we have

$$A \circ \gamma(w) = A(y) + H_{(v,\, b_2)}(W, W)(w^2/2) + \ldots$$

while

$$B \circ \gamma(w) = B(y) + H_{(y,b_1)}(W,W)(w^2/2) + \ldots.$$

Thus $H_{(y,b_2)}(W, W) \leq H_{(y,b_1)}(W, W)$ for all W, and if $H_{(y,b_1)}$ is negative definite on a subspace V then so is $H_{(y,b_2)}$, which implies $I_{(y,b_1)} \leq I_{(y,b_2)}$, and similarly, $P_{(y,b_1)} \geq P_{(y,b_2)}$.

If $g(t)$ is not a conjugate point of $g(0)$, then $H_{(y,t)}$ is non-singular on a neighborhood of (y, t), since the conjugate points are isolated, and hence,

(4) $\quad I_{(y,t)} \quad$ and $\quad P_{(y,t)}$

are constant on a neighborhood of (y, t).

We now use the properties (2), (3), and (4) to compute $I(L_y)$. Let $a_1,\ldots,a_s$ be the points on $[0, b)$ that are conjugate to 0. If $0 < t < a_1$ we know $P_{(y,t)} = N$, $I_{(y,t)} = 0$, and $N_{(y,t)} = 0$ by theorem 9 and property (4). At $t = a_1$, $N_{(y,a_1)} = \deg g(a_1)$, $I_{(y,a_1)} = 0$ by (2) since $I_{(y,t)} = 0$ for $t < a_1$, and hence $P_{(y,a_1)} = N - \deg g(a_1)$. If $a_1 < t < a_2$, and t near a_1, $P_{(y,t)} \geq P_{(y,a_1)}$ by (2) and $P_{(y,t)} \leq P_{(y,a_1)}$ by (3), hence $P_{(y,t)} = N - \deg g(a_1)$, $N_{(y,t)} = 0$, and $I_{(y,t)} = \deg g(a_1)$. The situation then remains unchanged for $a_1 < t < a_2$ by (4). For $t = a_2$, we repeat the above reasoning to compute $N_{(y,a_2)} = \deg g(a_2)$, $I_{(y,a_2)} = \deg g(a_1)$, and $P_{(y,a_2)} = N - \Sigma_{0\leq t\leq a_2}\deg g(t)$. Continuing the argument, we obtain $I(L_y) = I_{(y,b)} = \Sigma_{0\leq t<b}\deg g(t)$.//

Section 10.5. Completeness.

The theorem that follows gives useful criteria for a Riemannian manifold to be complete. The analytic case was first studied by Hopf-Rinow. The approach we give essentially follows de Rham[2].

THEOREM 15. *If M is a connected Hausdorff Riemannian manifold, then (a), (b), (c), and (d), stated below, are equivalent statements, and anyone of them implies (e).*

(a) *The exponential map is everywhere defined on $T(M)$.*

(b) *The manifold is complete with respect to its Riemannian metric.*

(c) *Bounded closed sets in M are compact.*

(d) *The closed balls $\overline{B}(m, r)$ are compact for one m in M and all $r > 0$.*

(e) *Any two points in M can be joined by a geodesic segment whose length equals the distance between the two points.*

Proof. The implications (d) $\Rightarrow$ (c) $\Rightarrow$ (b) $\Rightarrow$ (a) are all simple. We show (a) implies (d) and (e). Let m be a fixed point of M, let $B_r = B(m, r)$ $S_r = \overline{B}(m, r)$, and let $E_r = [p \text{ in } S_r\text{: there is a geodesic segment } \gamma \text{ from } m \text{ to } p \text{ with } |\gamma| = d(m, p)]$. We show E_r is compact and $E_r = S_r$ for all r, which proves (d) and (e).

LEMMA 1. *The set E_r is compact for all r.*

Proof. Fix r and let $[m_k]$ be a sequence of points in E_r. By (a) there exist points X_k in M_m such that $\exp_m X_k = m_k$ for all k. This follows since a geodesic can always be written as a composite map which is the exponential of a ray in a tangent space. Then $|X_k| < r$ for all k, hence $[X_k]$ is a sequence of points in the compact set $\overline{B}(0, r)$ in the Euclidean space M_m. Thus we obtain a subsequence (which we reindex if necessary) $[X_k]$ that converges to X in M_m with $|X| \leq r$. The corresponding subsequence $[m_k]$ converges to $\exp_m X$, which lies in E_r since $\exp_m$ is C^∞.

LEMMA 2. If $E_r = S_r$ for a fixed r, and $d(m, p) > r$, , then there is a point $\overline{m}$ such that $d(m, \overline{m}) = r$ and $d(m, p) = r + d(\overline{m}, p)$.

Proof. For each integer $k > 0$, let γ_k be a broken C^∞ curve from m to p with $|\gamma_k| < d(m, p) + (1/k)$. Let m_k be the last point on each γ_k that lies in S_r, so $d(m, m_k) = r$. Since S_r is compact, the sequence $[m_k]$ has a limit point $\overline{m}$ and $d(m, \overline{m}) = r$. But $d(m_k, p) \leq |\gamma_k|_{m_k}^p = |\gamma_k|_m^p - |\gamma_k|_m^{m_k} \leq |\gamma_k| - r < d(m,p) + (1/k) - r$. Hence $d(\overline{m}, p) \leq d(m, p) - r$, and the triangle inequality proves the opposite inequality.

LEMMA 3. *For $r \geq 0$, $E_r = S_r$.*

Proof. The proof uses a continuous induction argument on r. By definition, $E_r \subset S_r$ for all r. For $f = 0$, $E_0 = S_0$. If $E_r = S_r$, then trivial $E_{r'} = S_{r'}$ for all $r' < r$. Conversely, if $E_{r'} = S_{r'}$ for all $r' < r$, then $E_r = S_r$. This follows by taking any point p in S_r and then choosing $[p_k] \to p$ such that each p_k in some $S_{r'}$ for $r' < r$. Hence each p_k in $E_{r'} \subset E_r$, and E_r is compact, which implies the limit p is in E_r.

Finally, if $E_r = S_r$, then there is an $\epsilon > 0$ such that $E_{r+\varepsilon} = S_{r+\varepsilon}$. Since S_r is compact, we obtain a number $2\epsilon > 0$ such that for all p in S_r the map $\exp_p$ is a diffeo from $[X \text{ in } M_p\text{: } |X| < 2\epsilon]$ onto $B(p, 2\epsilon)$. Take p in $S_{r+\varepsilon}$. By lemma 2, there is a point $\bar{m}$ with $d(m, \bar{m}) = r$ and $d(\bar{m}, p) = d(m, p) - d(m, \bar{m}) \leq r + \epsilon - r \leq \epsilon$. Hence there is a geodesic segment γ_1 from m to $\bar{m}$ with $|\gamma_1| = r$, and a geodesic segment γ_2 from $\bar{m}$ to p with $|\gamma_2| = d(\bar{m}, p)$ Joining γ_1 and γ_2 gives a broken C^∞ curve γ from m to p with $|\gamma| = d(m, p)$. Parameterizing γ by arc length, there can be no breaks in γ, so γ is geodesic. Thus p in $E_{r+\varepsilon}$.//

We can now prove a classical theorem which illustrates how assumptions about the Riemannian curvature can affect the topology of a manifold.

THEOREM 16 (Bonnet). *If M is a complete connected Riemannian manifold with Riemannian curvature $\geq K > 0$, then M is compact and its diameter is $\leq \pi / \sqrt{K}$.*

Proof. We show on every geodesic g there is a conjugate point of $g(0)$ on $[0, \pi / \sqrt{K}]$. If m a fixed point of M, then by completeness every point p of M can be joined to m by a geodesic segment whose length is $d(p, m)$. By Theorem 11, this geodesic has no conjugate point of m before p, hence $d(m, p) \leq \pi / \sqrt{K}$.

Let g be a geodesic with unit tangent T, $g(0) = m$, and let e be a unit parallel field along g which is orthogonal to T. Let $W_t = (\sin \sqrt{K}t)e_t$. Then W is orthogonal to T, W vanishes at 0 and $\pi/\sqrt{K}$,

$$D_T W = (\sqrt{K} \cos \sqrt{K}t)e_t, \text{ and}$$

$$L''_W(0) = \int_0^{\pi/\sqrt{K}} [\langle R(W, T)W, T\rangle + \langle D_T W, D_T W\rangle] dt$$

$$= \int_0^{\pi/\sqrt{K}} [-K(t) \sin^2 \sqrt{K}t + K \cos^2 \sqrt{K}t]dt$$

$$\leq K \int_0^{\pi/\sqrt{K}} [\cos^2 \sqrt{K}t - \sin^2 \sqrt{K}t]dt = 0,$$

where $K(t) = \langle R(e, T)T, e\rangle$. If the interval $[0, \pi/\sqrt{K}]$ was free of conjugate points, then by lemma 2 of section 10.2, $L''_W(0) > L''_Z(0) = 0$, where $Z = 0$ is the unique Jacobi field along g, which coincides with W at 0 and $\pi/\sqrt{K}$. This contradiction proves the theorem.//

The following theorem, due to K. Nomizu and H. Ozeki, settles the question of the existence of complete Riemannian metrics on a paracompact (or Riemannian) manifold. A Riemannian metric is bounded if the manifold is bounded with respect to the induced metric function.

THEOREM 17. *Let M be a connected Hausdorff C^∞ manifold, If G is any Riemannian metric on M, then there exist Riemannian metrics G_1 and G_2, both conformal to G, with G_1 complete and G_2 bounded.*

Proof. Since there is more than one Riemannian metric involved, write $G_i(X, Y)$ rather than $\langle X, Y\rangle_i$ for the metric tensor applied to a pair of vectors, d_i for the metric, and $B_i(m, r)$ for the corresponding r-ball neighborhoods.

Using the metric G, for each p in M, let $r(p) = \sup\,[r\colon\ \overline{B}(p, r)$ is compact$]$. If $r(p) = \infty$ for some p, then G is complete by theorem 15. Suppose $r(p) < \infty$ for all p, and we construct G_1.

Notice $|r(p) - r(m)| \leq d(p, m)$ for all p and m, for if $r(p) > r(m) + d(p, m)$, one could increase $r(m)$; hence $r(p) \leq r(m) + d(p, m)$ for all p and m, and the inequality follows. This proves r is continuous.

Since M is paracompact, it is easy to show there is a real valued C^∞ function f on M with $f(p) > 1/r(p)$ for all p. Let $G_1(X, Y) = f^2(m)G(X, Y)$ for X, Y in M_m, which defines a C^∞ Riemannian metric G_1 on M.

That G_1 is complete will follow by showing $B(p, 1/3)$ is contained in $B(p, r(p)/2)$, and hence $\overline{B}_1(p, 1/6)$ is compact. This implies every Cauchy sequence in the G_1 metric must converge. To show this, take p in M and take m such that $d(p, m) \geq r(p)/2$. Let γ be a broken C^∞

curve from p to m, which is parameterized by G-arc length, i.e., if T is the tangent to γ, then $G(T, T) = 1$ and γ defined on $[0, L]$ where L is the G-length of γ, so $L \geq r(p)/2$. Letting L_1 be the G_1-length of γ, $L_1 = \int_0^L \sqrt{G_1(T, T)}dt = \int_0^L (f \circ \gamma)dt = f(\bar{p})L > L/r(\bar{p})$, where $\bar{p}$ is on γ between p and m. But $|r(\bar{p}) - r(p)| \leq d(p, \bar{p}) \leq L$; hence $r(\bar{p}) \leq r(p) + L$ and $L' > L/(r(p) + L) > L/3L = 1/3$. Hence $d_1(p, m) \geq 1/3$, so $B_1(p, 1/3) \subset B(p, r(p)/2)$.

For the second part of the theorem we may assume $G = G_1$ is complete. Fix a point m in M and let f be a real valued C^∞ function on M such that $f(p) > d(m, p)$ for all p. Let $G_2 = e^{-2f}G$, and we show G_2 is bounded. Take p in M and let γ be a geodesic from m to p with tangent T such that $G(T, T) = 1$, γ defined on $[0, L]$, and $L = d(m, p)$. Then $f \circ \gamma(t) > d(m, \gamma(t)) = t$ for all t. Letting L_2 be the G_2-length of γ, $L_2 = \int_0^L \sqrt{G_2(T, T)}dt = \int_0^L e^{-f}dt < \int_0^L e^{-t}dt < \int_0^\infty e^{-t}dt = 1$. Hence $d_2(m, p) < 1$ for all m and p.//

Corollary . Every Riemannian metric on a manifold is complete iff the manifold is compact.

For further work on completeness see the papers of J. A. Wolf[1 and 2] and P. A. Griffiths.

Section 10.6. *Manifolds with constant Riemannian curvature.*

THEOREM 18. *Let M and M' be connected Riemannian manifolds with M complete. Let f be an isometry of M into M'. Then f is onto, f is a covering map, and M' is complete.*

Proof. To show f is onto we show $f(M)$ is open (which is trivial since f is a local diffeo) and closed. Take m' in $\overline{f(M)}$, let B' be a convex neighborhood of m', let $p' = f(p)$ be in B', and let g' be the unique geodesic in B' from p' to m' with $g'(0) = p'$ and $g'(1) = m'$. Let g be the unique geodesic in M with $g(0) = p$ and $f_*T_g(0) = T_{g'}(0)$. Since f an isometry, $f \circ g$ is a geodesic in M', and by uniqueness, $f \circ g = g'$. Since M is complete, $g(1) = m$ is defined; hence $f(m) = m'$ and f is onto. We have also shown M' is complete.

It is trivial that f evenly covers, since f preserves locally convex neighborhoods; thus for m' we choose a convex neighborhood B', and $f^{-1}(B')$ is a union of disjoint convex nieghborhoods, each of which

f maps diffeomorphically onto B'.//

THEOREM 19. *Let M be a connected and simple connected, complete Riemannian manifold with constant Riemannian curvature K. Then M is isometric to Euclidian space, spherical space, or hyperbolic space, when $K = 0$, $K > 0$, or $K < 0$, respectively.*

Proof. Let g be a geodesic in M parameterized by arc length with $g(0) = m$. Let e be a parallel unit field along g, which is orthogonal to T, the unit tangent to g. Let $Z(t) = a(t)e(t)$ be a C^∞ field along g. Then $D_T Z = a'e$ and $D_T^2 Z = a''e$. Thus Z is a Jacobi field if $D_T^2 Z = R(T, Z)T$ or

$$\langle D_T^2 Z, Z\rangle = \langle R(T, Z)T, Z\rangle = -K\langle Z, Z\rangle, \text{ i.e., } a''a = -Ka^2$$

or $a'' + Ka = 0$. This differential equation has solutions uniquely determined by $a(0)$ and $a'(0)$. If $a(0) = 0$, then $Z(t) = (\exp_m)_* tA'$ where A' is the constant field on M_m with $A' = a'(0)e$. This equality follows from the fact that the right side is a Jacobi field and a Jacobi field is determined by Z and $D_T Z$ at one point. Hence $\langle Z, Z\rangle = t^2\langle \exp_* A'(t), \exp_* A'(t)\rangle = a^2(t)$.

When $K = 0$, then $a'' = 0$ and $a = ct$ where $c = a'(0)$. Thus $\langle \exp_* A'(t), \exp_* A'(t)\rangle = c^2 = \langle A'(t), A'(t)\rangle$, and $\exp_m$ is an isometry from M_m onto M. Apply the previous theorem to obtain $\exp_m$ is a covering map. Since M is simply connected, $\exp_m$ is a diffeo, hence M is isometric to M_m, and M_m is trivially isometric to Euclidean space.

When $K < 0$, let M' be hyperbolic space for $K < 0$ (section 6.7). We know $\exp_0$: $M'_0 \to M'$ is a diffeo so let $E = (\exp_0)^{-1}$. Choose an orthonormal base $e'_1,\ldots,e'_n$ of M'_0 and an orthonormal base $e_1,\ldots,e_n$ of M_m, where m an arbitrary point of M. Let F: $M'_0 \to M_m$ by $F(e'_1) = e_i$. Let f: $M' \to M$ by $f = \exp_m \circ F \circ E$. Then $f_* Z'(t) = Z(t)$ along corresponding geodesics in M' and M, and $\langle f_* Z', f_* Z'\rangle = a^2(t) = \langle Z', Z'\rangle$. Thus f_* is an isometry. Now apply previous theorem to obtain f a diffeo.

When $K > 0$, then $Z = (\sin\sqrt{K}t)e$ is a Jacobi field along any geodesic emanating from m (a fixed point in M). Thus every ray in M_m has a conjugate point at $\pi/\sqrt{K}$ units from the origin and $(\exp_m)_*$ has an $(n - 1)$ dimensional kernel at these points. Let $C = [X \text{ in } M_m\colon |X| = \pi/\sqrt{K}]$. Then $\exp_m|_C$ is completely singular and hence is a constant map since C is connected. From the nature of the Jacobi equations in the first paragraph there are no conjugate points in $B = [X \text{ in } M_m\colon\ |X| < \pi/\sqrt{K}]$. Now let M' be spherical space of curvature K, let p be any point in M'. We know $\exp_p$ is a diffeo on the set B' (corresponding to B) in M'_p. Define E and F as in the above paragraph (E defined on $B(p, \pi/\sqrt{K})$, the open ball), and let $f = \exp_m \circ F \circ E$ on $B(p, \pi/\sqrt{K})$ while $f(-p) = \exp_m(C)$. As in the above paragraph, f is an isometry on $B(p, \pi/\sqrt{K})$. Note what should be f_* at $-p$ is well-defined via the tangents to incoming geodesics. Thus we may define a map $g\colon\ B(-p, \pi/\sqrt{K}) \to M$ with $g(-p) = f(-p)$ and g_* at $-p$ determined by f_*. Then $f = g$ on their common domain and g is C^∞ and metric preserving at $-p$. Hence f is an isometry of M' onto M, and by the previous theorem f is a diffeo.//

Corollary. Let M and M' be the Riemannian manifolds, let $b = [e_1, \ldots, e_n]$ be an orthonormal base at m in M, and similarly, let b' be such a base at m' in M'. Let $F\colon\ M_m \to M_{m'}$ by $F(e_i) = e_i'$. The map F induces a correspondence between geodesics emanating from m and m', respectively, and also a correspondence between plane sections P and P' along these geodesics, via parallel translation of corresponding plane sections at m and m'. Thus for a geodesic g in M with $g(0) = m$, let g' be the geodesic in M' with $g'(0) = m'$ and $T_{g'}(0) = F(T_g(0))$; and for a plane section P in M_m let $P(t)$ be the parallel translate of P along g to $g(t)$, let $P' = F(P)$ and $P'(t)$ is the parallel translate of P' along g'. Suppose $K'(P'(t)) = K(P(t))$ for all geodesics and all plane sections (emanating from m and m'). Then there are neighborhoods B and B' of m and m', respectively, and a map $f\colon\ B \to B'$ which is an isometry (and a diffeo). Thus M and M' are locally isometric at m and m'.

Proof. Choose an $r > 0$ such that $\exp_m$ is a diffeo from $B(0, r)$ in M_m onto $B = B(m, r)$ in M and $\exp_{m'}$ is also a diffeo from $B'(0, r)$ in $M'_{m'}$ onto $B' = B'(m', r)$ in M'. Let $f = \exp_m \circ F \circ (\exp_{m'})^{-1}$ on B', so f is a diffeo. By the method of proof in the preceding theorem, f is an isometry.//

If in the above corollary we add the hypothesis that M and M' be complete, connected, and simply-connected, then it is an open question whether M is isometric to M'. When the Riemannian curvature is preserved for corresponding plane sections on once-broken geodesics, then Ambrose[1] has proven M is isometric to M'.

Section 10.7. Manifolds without conjugate points.

Most of the results of the next two sections are based on a paper by A. Preissmann and some informal notes by W. B. Houston, Jr.

Throughout this section let M be a complete connected Hausdorff Riemannian n-manifold. If m is a point of M and there exists no point of M that is conjugate to m, then m is called a *pole*.

THEOREM 20. *If m is a pole in M, then* $\exp_m$: $M_m \to M$ *is a covering map. Thus the simply connected covering of M is diffeo to R^n, and if M is simply connected, then M is diffeo to R^n.*

Proof. Letting $E = \exp_m$, we know E is onto since M is complete, and E is a local diffeo since m has no conjugate points. The metric tensor G of M induces a Euclidean metric on M_m whose distance function we denote by d. On the other hand, by requiring E to be an isometry, we define a metric tensor G_1 on M_m whose distance function we denote by d_1. The rays in M_m, emanating from the origin, are G_1-geodesics since E is connexion preserving. We now show these rays are minimizing G_1-geodesics from the origin.

Take any X in M_m, and let γ be a C^∞ curve from 0 to X with $\gamma(t)$ in $\bar{B}(0, |X|)$ for all t (B is the Euclidean ball). Assume γ parameterized so $|\gamma(t)| = t$, thus γ defined on $[0, |X|]$. Let T be the tangent to γ, then $T_t = R_t + V_t$ where R is the unit (outward) radial vector field on M_m (and $R_0 = T_0$), and V_t is orthogonal to R_t at each point. Computing the G_1-length of T, $|T|_1 = |E_*(R+V)| \geq |E_*(R)| = 1$ by the perpendicular lemma. Hence, $|\gamma|_1 = \int_0^{|X|} |T|_1 dt \geq |X|$, which implies $d_1(0, X) = |X|$, since the ray from 0 to X has G_1-length equal to $|X|$. Thus $\bar{B}_1(0, b) = \bar{B}(0, b)$ for all $b \geq 0$, and since the latter is compact so is the former. By the completeness theorem (15), M_m is complete with respect to the G_1-metric. By theorem 18, the map E: $M_m \to M$ is a covering map.//

Corollary. If M has non-positive Riemannian curvature, then all points are poles and R^n is simply connected covering space of M.

We now define the *universal covering manifold* $\bar{M}$, based at a point m in M, in a standard way. Let $\bar{M}$ be the set of equivalence classes of C^0-homotopic C^0-curves f defined on a finite interval such that $f(0) = m$ (see Hocking-Young, p. 188). Let $[f]$ denote the equivalence class of a curve f, and let $\pi: \bar{M} \to M$ denote the covering map where $\pi([f])$ is the endpoint of f. Define a C^∞ structure on $\bar{M}$ by demanding π to be a C^∞ map, and if M is Riemannian, define a Riemannian metric on $\bar{M}$ such that π is an isometry. We use repeatedly the fact that a C^0-curve f in M has a unique lifting $\bar{f}$ in $\bar{M}$ such that $\pi \circ \bar{f} = f$ once one has prescribed $\bar{f}(0)$. Let $f \sim h$ denote the fact that f is homotopic to h under a fixed end-point homotopy, and let $\bar{m}$ by the constant path at m.

THEOREM 21. *Let f be a finite curve in M and let $b = \inf\,[|h|$: h is a broken C^∞ curve and $h \sim f]$. Then there exists a geodesic g such that $g \sim f$ and $|g| = b$. Thus in every homotopy class of curves (with fixed end-points) there is a geodesic whose length is the absolute minimum for the lengths of all broken C^∞ curves in the homotopy class.*

Proof. Let $\bar{M}$ be the universal covering manifold based at $m = f(0)$. Since M is complete, $\bar{M}$ is complete, and hence there exists a geodesic $\bar{g}$ from $[\bar{m}]$ to $[f]$ which gives the distance in $\bar{M}$ between these two points. Then $g = \pi \circ \bar{g}$ is a geodesic in M since π is an isometry, and $g \sim f$ since $\bar{M}$ is simply connected. If h is a broken C^∞ curve with $h \sim f$, then lift h to a curve $\bar{h}$ starting at $[\bar{m}]$ and obtain a broken C^∞ curve $\bar{h}$ from $[\bar{m}]$ to $[f]$. Since $\bar{g}$ gives the distance, $|\bar{h}| \geq |\bar{g}| = |g|$, thus $|g| = b$.//

THEOREM 22. *Let m be a pole in M and let g_1 and g_2 be geodesics emanating from m that intersect later. If $g_1 \sim g_2$, then $g_1 = g_2$ (when both parameterized by arc length).*

Proof. Let $\bar{M}$ be the universal covering manifold based at m with π an isometry. Let $\overline{\exp}$: $M_m \to \bar{M}$ by $\overline{\exp}\,(X) = [\exp_m tX\colon\ 0 \leq t \leq 1]$. Then $\pi \circ \overline{\exp} = \exp_m$ and $\overline{\exp}$ is C^∞, since locally $\overline{\exp} = \pi^{-1} \circ \exp_m$. Moreover, $\overline{\exp}$ is an isometry, for m is a pole. Since M_m is simply connected, $\overline{\exp}$ is a diffeo by theorem 18. If $g_i(t) = \exp tX_i$ where $g_1(1) = g_2(1)$, and $g_1 \sim g_2$, then $[g_1] = [g_2]$. Since $\overline{\exp}$ is a diffeo, this implies $X_1 = X_2$, which implies $g_1 = g_2$.//

We remark that one can always define the C^∞ map $\overline{\exp}$: $M_m \to \bar{M}$ (base point m) with $\pi \circ \overline{\exp} = \exp_m$. The map $\overline{\exp}$ will be onto if M

is complete, but it will not in general be locally one-to-one.

Corollary 1. If m is a pole in M and M is simply connected, then for any point p in M there is a unique geodesics through m and p.

Corollary 2. If M is simply connected and has only non-positive Riemannian curvature, then there is a unique geodesic through any two points of M.

Section 10.8. *Manifolds with non-positive curvature.*

We add to the standard hypothesis of the last section the assumption that $K(P) \leq 0$ for all plane sections P of M.

LEMMA 1. *Let f be a finite curve in M parameterized by arc length, and let m be a point of M. Let $\bar{f}$ be any lifting of f to the covering space M_m (see theorem 20). Then $|f| \geq |\bar{f}|$, the Euclidean length of $\bar{f}$ in M_m. If $K < 0$, then $|f| > |\bar{f}|$ unless $\bar{f}$ is a segment of a ray emanating from zero in M_m.*

Proof. By theorem 13, if T is a vector tangent to M_m, then $|(\exp_m)_* T| \geq |T|$. If $K < 0$, then $|(\exp_m)_* T| > |T|$ unless T is a radial vector tangent to a ray through zero.//

THEOREM 23. *Let p_1, p_2, and p_3 be distinct points of M which are joined by geodesics g_1, g_2, and g_3 where g_1 joins p_2 and p_3, etc., (see Fig. 10.6). Assume the three points are not on one geodesic and the broken loop formed by the three curves is homotopic to zero. Let θ_i be the unique angle at p_i made by the intersecting geodesics with $0 < \theta_i < \pi$. Then*

$$|g_1|^2 \geq |g_2|^2 + |g_3|^2 - |g_2|\,|g_3| \cos \theta_1, \quad \text{and} \quad \theta_1 + \theta_2 + \theta_3 < \pi.$$

If $K < 0$ on M, these inequalities are strict.

Proof. Let $m = p_1$ and let $\bar{g}_2$ and $\bar{g}_3$ be the rays through zero in M_m such that $\exp_m \circ \bar{g}_i = g_i$ for $i = 2, 3$. Let X_2 and X_3 be the end-points of $\bar{g}_3$ and $\bar{g}_2$, respectively. Since the loop formed by g_2, g_1, and g_3 is homotopic to zero, we can lift g_1 to a curve $\bar{g}_1$ joining X_2 and X_3. By the preceding lemma, $|g_1| \geq |\bar{g}_1| \geq d(X_2, X_3)$, where d is the Euclidean distance in M_m. By the law of cosines in M_m, $(d(X_2, X_3))^2 = |g_2|^2 + |g_3|^2 - |g_2|\,|g_3| \cos \theta_1$, which proves the first inequality.

For the second inequality, we construct a triangle in R^2 whose sides have lengths $a_i = |g_i|$ and label the angles at the appropriate corners by ϕ_i. Then $(a_1)^2 = (a_2)^2 + (a_3)^2 - a_2 a_3 \cos \phi_1$; hence $\cos \theta_1 \geq \cos \phi_1$ and $\theta_1 \leq \phi_1$. Similarly, $\theta_i \leq \phi_i$ for all i, and $\theta_1 + \theta_2 + \theta_3 \leq \phi_1 + \phi_2 + \phi_3 = \pi$.

If $K < 0$, then $|g_1| > |\bar{g}_1|$ and the strict inequalities then follow.//

Corollary 1. The sum of the interior angles $(0 < \theta_i < \pi)$ of a geodesic quadralateral which is homotopic to zero is $\leq 2\pi$. If $K < 0$, then the sum is $<2\pi$.

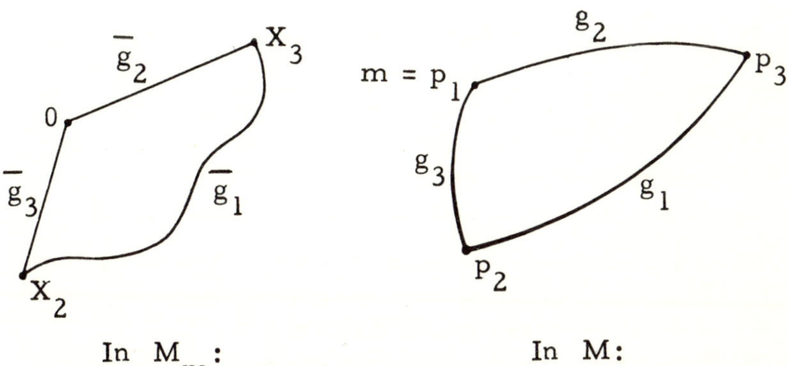

Fig. 10.6 Geodesic Triangle

Corollary 2. Let m be in M, and let g be a geodesic that does not pass through m. Then there cannot be two distinct geodesics g_1 and g_2 from m to g which intersect g orthogonally such that the geodesic triangle formed is homotopic to zero.

Proof. The sum of the interior angles of the geodesic triangle would be greater than π.//

Corollary 3. Let M be simple connected, m in M, and g a geodesic that does not pass through m. Then there is a unique geodesic f from m to g which is orthogonal to g and $|f| \leq d(m, g(t))$ for all t.

Proof. Let f_t be the unique geodesic from m to $g(t)$, let $L(t) = |f_t| = d(m, g(t))$, and let g_t be g restricted to the interval $[0, t]$ or $[t, 0]$, as the case may be. Let θ be the angle between f_0 and g_t for

$t > 0$. We show that $L(t) \to \infty$ as $t \to +\infty$ or $-\infty$. For $t > 0$,

$$L^2(t) = |f_t|^2 \geq |f_0|^2 + |g_t|^2 - |f_0|\,|g_t| \cos\theta =$$

$$= |f_0|^2 + |g_t|(|g_t| - |f_0| \cos\theta).$$

As $t \to \infty$, $|g_t| \to \infty$, and hence $L(t) \to \infty$. Similarly, $L(t) \to \infty$ as $t \to -\infty$.

By theorem 7, a point t' is a critical point of L if and only if $f_{t'}$ is orthogonal to g. By corollary 2 there can be at most one critical point of L, and that must be an absolute minimum by the first paragraph.//

For further results see Preissman and Helgason.

Problems

93. Using the notation of section 3.4, show that T_v is a Jacobi field on a surface of revolution. If $G = \langle T_v, T_v \rangle$ and S is arc length along the meridians, show $d^2\sqrt{G}/ds^2 = -K\sqrt{G}$.

94. If M is a complete Riemannian 2-manifold, show the locus of first (those nearest the origin on each ray) conjugate points in M_m is a C^∞ curve (see S. B. Myers).

95. Show the Hessian is well-defined, symmetric, and bilinear.

96. If d is the function defined in the lemma in section 10.4, show d is C^∞ on C.

97. If $M = R^3$, g is the x-axis from $(a, 0, 0)$ to $(b, 0, 0)$ with $a < b$, and C_1 and C_2 are the planes $x = a$ and $x = b$, respectively, check the lemma in section 10.4.

98. A submanifold V of a manifold M is *totally geodesic* with respect to a connexion D if any geodesic that is tangent to V at a point lies wholly in V. If V and W are compact totally geodesic submanifolds, of dimension r and s, respectively, lying in a Riemannian n-manifold M of positive Riemannian curvature and $r + s \geq n$, show $V \cap W$ is non-empty (see Frankel).

99. Find a condition relating curvature and parallel translation that will insure the existence of complete totally geodesic

submanifolds in a Riemannian manifold (see Hermann or Helgason).

100. If M is an oriented n-manifold and α is a $C^{\infty}(n-1)$-form on M with compact support, show $\int_M d\alpha = 0$ (see Nijenhuis and Richardson).

BIBLIOGRAPHY

Abraham, R.
Piecewise differentiable manifolds and the space-time of general relativity. J. Math. and Mech. 11 (1962), 553-592.

Adler, A.
Characteristic classes of homogeneous spaces. Trans. A. M. S. 86 (1957), 348-365.

Ahlfors, L. V. and Sario, L.
"Riemann Surfaces". Princeton U. Press, 1960.

Allendoerfer, C. B.
1. The Euler number of a Riemann manifold. Amer. J. Math 62 (1940), 243-248.
2. Characteristic cohomology classes in a Riemann manifold. Ann. of Math. 51 (1950), 551-570.

Allendoerfer, C. B., and Weil, A.
The Gauss-Bonnet theorem for Riemannian polyhedra. Trans. A. M. S. 53 (1943), 101-129.

Alexandroff, P., und Hopf, H.
"Topologie". I. Springer, Berlin, 1935. Reprinted by Edwards Brothers, Ann Arbor, Michigan, 1945.

Ambrose, W.
1. Parallel translation of Riemannian curvature. Ann. of Math. 64 (1956), 337-363.
2. The index theorem in Riemannian geometry. Ann. of Math. 73 (1961), 49-86.
3. The Cartan structural equations in classical Riemannian geometry. J. Ind. M. S. 24 (1960), 23-76.

Ambrose, W., and Singer, I. M.
A theorem on holonomy. Trans. A. M. S. 75 (1953), 428-443.

Ambrose, W., Palais, R. S., and Singer, I. M.
Sprays. Anais Acad. Bras. Cincencias 32 (1960), 163-178.

Auslander, L., and Mackenzie, R. E.
"Introduction to Differentiable Manifolds". McGraw-Hill, 1963.

Bol, G.
Uber Nabelpunkte auf einer Eiflache, Math. Z. 49 (1944), 389-410.

Bonnesen, T. and Fenchel, W.
"Theorie der Kovexen Körper". Springer, Berlin, 1934.

Bott, R.
Morse theory and its application to homotopy theory. Lecture notes, U. of Bonn, 1960.

Buck, R. C.
"Advanced Calculus". McGraw-Hill, 1956.

Cairns, S. C.
A simple triangulation method for smooth manifold, Bull, A. M. S. 67 (1961), 389-390.

Cartan, E.
"Lecons sur la Geometrie des Espaces de Riemann". Gauthier-Villars, Paris, 1928, 2nd ed., 1946.

Cartan, H.
La transgression dans un groupe de Lie et dans un espace fibre principal.

Colloque de Topologie (Espaces Fibres), Bruxelles, 1950.

Chern, S. S.
1. Topics in differential geometry. Inst. Adv. Study, Princeton, 1951.
2. Differentiable manifolds. Lecture notes, U. of Chicago, 1959.
3. On the curvatura integra in a Riemannian manifold. Ann. of Math. 46 (1945), 674-684.

Chevalley, C.
1. "Theory of Lie Groups". vol. 1, Princeton U. Press, 1946.
2. "Fundamental Concepts of Algebra", Academic Press, 1956.

Coddington, E. A., and Levinson, N.
"Theory of Ordinary Differential Equations", McGraw-Hill, 1955.

Crittenden, R., and Bishop, R.
"Geometry of Manifolds", Academic Press, 1962.

DeRham, G.
1. "Varietes Differentiables". Hermann, Paris, 1955.
2. Sur la reductibilite d'un espace de Riemann. Comm. Math Helv. 26 (1952), 328-344.

Ehresmann, C.
Les connexions infinitesimales dans un espace fibre differentiable. Colloque de Topologie (Espaces Fibres), Bruxelles, 1950.

Eisenhart, L. P.
1. "An Introduction to Differential Geometry". Princeton U. Press, 1947.
2. "Riemannian Geometry". Princeton U. Press, 1949.

Fenchel, W.
On total curvature of Riemannian manifolds. London Math. Soc. 15 (1940), 15-22.

Frankel, T.
Manifolds with positive curvature. Pac. J. Math. 11 (1961), 165-174.

Goldberg, S. I.
"Curvature and Homology," Academic Press, 1962.

Griffin, Jr., J. S.
On the rotation number of a normal curve. Comp. Math. 13 (1958), 270-276.

Guggenheimer, H.
"Differential Geometry". McGraw-Hill, 1963.

Halmos, P. R.
"Finite-Dimensional Vector Spaces". D. Van Nostrand, 1958.

Hamburger, H. L.
Beweis einer Caratheodoryschen Vermutung. Acta Math. 73 (1941).

Helgason, S.
"Differential Geometry and Symmetric Spaces". Academic Press, 1962.

Hermann, R.
Existence in the large of totally geodesic submanifolds of Riemannian spaces. Bull. A. M. S. 66 (1960), 59-61.

Hicks, N. J.
1. Submanifolds of Semi-Riemannian manifolds. Rend. Cir. Mat. Palermo 12 (1964), 1-13.
2. Connection preserving, conformal, and parallel maps. Mich. Math. J. 10 (1963), 295-302.

Hilbert, D., and Cohn-Vossen, S.
"Geometry and the Imagination." Chelsea, 1952.

Hirsch, M.
On imbedding differentiable manifolds in Euclidean space. Arn. of Math. 73 (1961), 566-571.

Hocking, J. G., and Young, G. S.
"Topology." Addison-Wesley, 1961.

Hopf, H.
1. Uber die Curvatura integra geschlossener Hyperflachen. Math. Ann. 95 (1925), 340-367.
2. Uber die Drehung der Tangenten und Sehnen ebener Kurven, Comp. Math. 2 (1935), 50-62.
3. Uber Flachen mit einer Relation zwischen den Hauptkrummungen. Math. Nachr. 4 (1951), 232-249.

Jacobson, N.
"Lectures in Abstract Algebra," II. D. Van Nostrand, 1953.

Johnson, H. H.
Terminating prolongation procedures. Pac. J. Math. 10 (1960), 577-583.

Kaplan, W.
"Advanced Calculus." Addison-Wesley, 1952.

Kervaire, M.
A manifold which does not admit any differentiable structure. Comment. Math. Helv. 34 (1960), 257-270.

Kelley, J. L.
"General Topology." D. Van Nostrand, 1955.

Klingenberg, W.
"Riemannsche Geometrie Im Grossen." Math, Inst., U. of Bonn, 1962.

Kobayashi, S., and Nomizu, K.
"Foundations of Differential Geometry." John Wiley and Sons, 1963.

Kuranishi,
On E. Cartan's prolongation theorem of exterior differential systems. Amer. J. Math. 79 (1957), 1-47.

Lang, S.
"Introduction to Differentiable Manifolds." Interscience, 1962.

Levi-Civitae, T.
"The Absolute Differential Calculus." Blackie, London, 1927.

Mackey, G. W.
"Mathematical Fundations of Quantum Mechanics." W. A. Benjamin, 1963.

Massey, W. S.
Surfaces of Gaussian curvature zero in Euclidean 3 space. Tohoku Math. J. 14 (1962).

Meyers, S. B.
Connections between differential geometry and topology. Duke Math. J. 1 (1935), 376-391.

Miller, K. S., and Murray, F. J.
"Existence Theorems for Ordinary Equations." New York U. Press, 1954.

Milnor, J.
1. Differential Topology. Lecture notes, Princeton U., 1958
2. On manifolds homeomorphic to the 7-sphere. Ann. of Math. 64 (1956), 399-405.
3. "Morse Theory." Princeton U. Press, 1963.

Morse, M.
"The Calculus of Variations in the Large." Amer. Math. Soc. Coll. Pub., 1934.

Munkres, J. R.
1. "Elementary Differential Topology." Princeton U. Press, 1963.
2. Obstructions to the smoothing of piecewise-differentiable homeomorphisms. Ann. of Math., 72 (1960), 521-554.

Nomizu, K.
1. "Lie groups and Differential Geometry." Math. Soc. of Japan, 1956.
2. Invariant affine connexions on homogeneous spaces. Amer. J. Math. 76 (1954), 33-65.

Newns, W. F. and Walker, A. G.
Tangent planes to a differentiable manifold. J. London Math. S. 31 (1956), 400-407.

Nickerson, H. K., Steenrod, N. E., and Spencer, D. C.
"Advanced Calculus." D. Van Nostrand, 1959.

Nijenhuis, A., and Richardson, R. W.
A theorem on maps with non-negative Jacobians. Mich. Math. J. 9 (1962), 173-176.

Palais, R. S.
1. A definition of the exterior derivation in terms of Lie derivatives. Proc. A. M. S. 45 (1954), 902-908.
2. A global formulation of the Lie theory of transformation groups. Thesis, Harvard University, 1956.
3. Lectures on Morse theory. Lecture notes, Brandeis U., 1963.
4. Natural operations on differential forms. Trans. A. M. S. 92 (1959), 125-141.

Pervin, W. J.
"Foundations of General Topology." Academic Press, 1964.

Pohl, W. F.
Differential geometry of higher order. Topology 1 (1962), 169-211.

Pogorelov, A. V.
A new proof of rigidity of convex polyhedra. Uspehi Mat. Nauk. (N. S.) 11 (1956), 207-208.

Preissmann, A.
Quelque proprietes globables des espaces de Riemann. Comm. Mat. Helv. 15 (1943), 175-216.

Samelson, H.
Differential geometry. Lecture notes, U. of Michigan, 1955.

Seifert, H., and Threfall, W.
"Variationsrechnung im Grossen," Chelsea, 1951.

Smale, S.
1. Differential and combinatorial structures on manifolds. Ann. of Math. 74 (1961), 498-502.
2. A survey of some recent developments in differential topology. Bull. A. M. S. 69 (1963), 131-145.

Steenrod, N.
"Topology of Fibre Bundles." Princeton U. Press, 1951.

Sternberg, S.
"Lectures on Differential Geometry." Prentice-Hall, 1964.

Struik, D. J.
"Lectures on Classical Differential Geometry." Addison-Wesley, 1950.

Synge, J. L.
The first and second variations of length in Riemannian space. Proc. London Math. Soc. 25 (1926), 247-264.

Titus, C. J.
A theory of normal curves and some applications. Pac. J. Math. 10 (1960), 1083-1096.

Veblen, O.
"Invariants of Quadratic Differential Forms." Cambridge U. Press, 1927.

Veblen, O. and Whitehead, J. H. C.
"The Foundations of Differential Geometry." Cambridge U. Press, 1932.

Weil, A.
Sur les theoremes de Rham. Comm. Math. Helv. 26 (1952), 119-145.

Weyl, H.
"Space, Time, and Matter." Dover, 1922.

Whitehead, J. H. C.
1. Convex regions in the geometry of paths. Quart. J. Math. 3 (1932), 33-42.
2. Manifolds with transverse fields in Euclidean space. Ann. of Math. 73 (1961), 154-212.

Whitney, H.
1. On regular closed curves in the plane. Comp. Math 4 (1937), 276-284.
2. "Geometric Integration Theory." Princeton U. Press, 1957.

Willmore, T. J.
"An Introduction to Differential Geometry." Oxford, 1959.

INDEX